Original Title:
BASIC MATHEMATICS
A Comprehensive Guide to Fundamental Concepts for Students and Professionals

Author:
MBA. Luque Zevallos Helbert Justo

Independently published

Available at https://www.amazon.com

Total or partial reproduction of this work is prohibited.

ALL RIGHTS RESERVED

Language: English

2024

First Edition

ISBN: 9798301465598

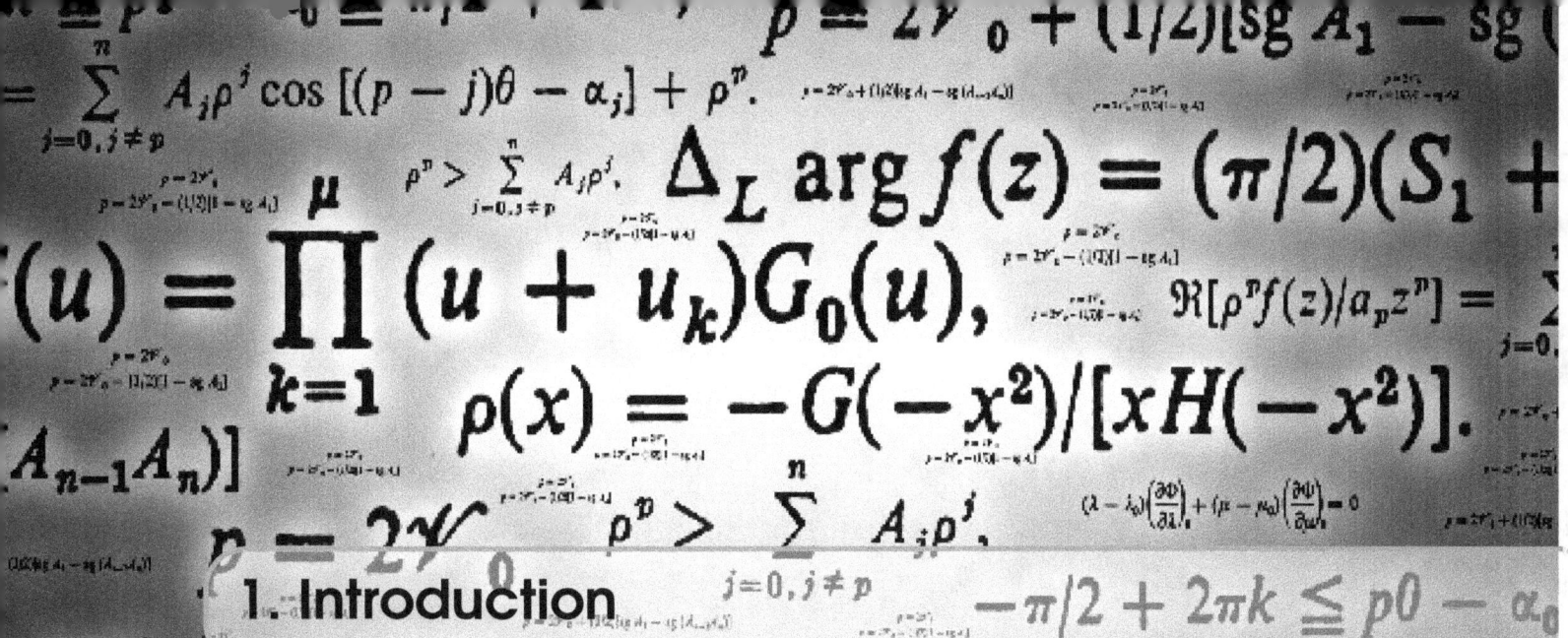

1. Introduction

This book is a comprehensive introduction to the fundamental concepts of mathematics, specifically designed for students and professionals seeking a solid foundation in essential topics ranging from algebra to calculus and beyond.

The book *Basic Mathematics*, written by Helbert Justo Luque Zevallos, offers a detailed and rigorous study of the fundamental mathematical principles essential for both students beginning their university studies and professionals seeking a comprehensive review. Covering topics such as the real number system, basic operations, properties of numbers, and more advanced subjects like equations, inequalities, vectors, matrices, and systems of equations, this text provides the tools needed to understand and apply mathematics in various scientific and technological contexts.

Each chapter begins with a clear theoretical introduction, followed by solved examples that demonstrate the application of the discussed concepts. The exercises proposed at the end of each section are designed to reinforce learning and encourage independent reasoning. Additionally, the book introduces practical applications that link mathematical theory to real-world situations, thereby preparing students for its use in professional and academic fields.

Basic Mathematics stands out for its focus on the clarity and depth of the material, ensuring that readers not only learn but also deeply understand the mathematical foundations that will form the basis of their future studies and professional careers. With its meticulously organized structure and accessible content, this book is an invaluable resource for anyone aspiring to master the fundamentals of mathematics.

MBA. Helbert Justo Luque Zevallos

Índice general

1 Introduction 3

2 Summary 11

3 Introduction 13

Mathematical Fundamentals and Algebra

1 Real Number System 17

1.1 Properties of Real Numbers — 17
- 1.1.1 Closure under Addition and Multiplication 17
- 1.1.2 Distributive Property 19
- 1.1.3 Commutative and Associative Properties 20

1.2 Intervals and Absolute Value — 22
- 1.2.1 Definition and Graphical Representation of Intervals 22
- 1.2.2 Types of Intervals (Open, Closed, Half-Open) 24
- 1.2.3 Calculation and Properties of Absolute Value 25

1.3 Operations with Real Numbers — 27
- 1.3.1 Addition and Subtraction of Real Numbers 27
- 1.3.2 Multiplication and Division of Real Numbers 29
- 1.3.3 Powers and Roots of Real Numbers 31

1.4 Solved Exercises — 34

1.5 Proposed Exercises — 35
- 1.5.1 Properties of Real Numbers 35
- 1.5.2 Intervals and Absolute Value 35

| | 1.5.3 | Operations with Real Numbers | 36 |

2 Equations and Inequalities … 37

2.1 Solving Polynomial Equations — 37
	2.1.1	First-Degree Equations	37
	2.1.2	Second-Degree Equations	39
	2.1.3	Higher-Degree Equations	40

2.2 Equations with Radicals and Rationalization — 42
	2.2.1	Definition of Radicals	42
	2.2.2	Simplification and Rationalization of Radicals	44
	2.2.3	Solving Radical Equations	46

2.3 Inequalities with Absolute Value and Their Graphical Representation — 47
	2.3.1	Definition of Absolute Value Inequalities	47
	2.3.2	Step-by-Step Solution of Inequalities	48
	2.3.3	Graphical Representation of Solutions	49

2.4 Solved Exercises — 51

2.5 Proposed Exercises — 54
	2.5.1	Solving Polynomial Equations	54
	2.5.2	Equations with Radicals and Rationalization	54
	2.5.3	Inequalities with Absolute Value and Their Graphical Representation	54

3 Complex Number System … 57

3.1 Representation of Complex Numbers in the Plane — 57
	3.1.1	Rectangular Form of a Complex Number	57
	3.1.2	Representation in the Argand Plane	59
	3.1.3	Modulus and Argument of a Complex Number	62

3.2 Operations with Complex Numbers — 65
	3.2.1	Addition and Subtraction of Complex Numbers	65
	3.2.2	Multiplication and Division in Rectangular Form	67
	3.2.3	Conjugate of a Complex Number	69

3.3 De Moivre's Formula and Its Applications — 72
	3.3.1	Definition of De Moivre's Formula	72
	3.3.2	Powers of Complex Numbers Using De Moivre's Formula	74
	3.3.3	Nth Roots of a Complex Number	76

3.4 Solved Exercises — 78

3.5 Proposed Exercises — 81
	3.5.1	Representation of Complex Numbers in the Plane	81
	3.5.2	Operations with Complex Numbers	82
	3.5.3	De Moivre's Formula and Applications	82

4 Vectors in $\mathbb{R}^2$ and $\mathbb{R}^3$ … 83

4.1 Definition and Operations with Vectors — 83
	4.1.1	Definition of Vectors	83
	4.1.2	Addition and Subtraction of Vectors	84
	4.1.3	Scalar Multiplication	88

4.2 Dot Product and Cross Product — 90
- 4.2.1 Definition of the Dot Product — 90
- 4.2.2 Properties of the Dot Product — 92
- 4.2.3 Definition of the Cross Product — 94

4.3 Parametric Equations and Cartesian Equations of Lines and Planes — 96
- 4.3.1 Parametric Equations of a Line — 96
- 4.3.2 Cartesian Equations of a Line — 97
- 4.3.3 Equation of a Plane — 99

4.4 Solved Exercises — 100

4.5 Proposed Exercises — 101
- 4.5.1 Vector Definitions and Operations — 101
- 4.5.2 Dot Product and Cross Product — 101
- 4.5.3 Parametric and Cartesian Equations of Lines and Planes — 102

II Analytic Geometry and Functions

5 Equations of Lines and Planes — 105

5.1 General Equation and Parametric Equation of a Line — 105
- 5.1.1 General Equation of a Line in the Plane — 105
- 5.1.2 Parametric and Vector Equations — 108
- 5.1.3 Conversion Between Forms of the Equation — 110

5.2 Conditions for Parallelism and Perpendicularity — 112
- 5.2.1 Parallel and Perpendicular Vectors — 112
- 5.2.2 Calculating the Slope to Determine Parallelism — 113
- 5.2.3 Applications in Solving Geometric Problems — 115

5.3 Equation of a Plane in Space — 117
- 5.3.1 Definition and Forms of the Plane Equation — 117
- 5.3.2 Parallel and Perpendicular Planes — 119
- 5.3.3 Intersection of Planes — 121

5.4 Solved Exercises — 123

5.5 Proposed Exercises — 125
- 5.5.1 General and Parametric Equations of a Line — 125
- 5.5.2 Conditions of Parallelism and Perpendicularity — 125
- 5.5.3 Equation of the Plane in Space — 125

6 Equations of Conics — 127

6.1 Circle: Equation and Properties — 127
- 6.1.1 Equation of the Circle in the Plane — 127
- 6.1.2 Center and Radius of a Circle — 129
- 6.1.3 Applications of the Circle in Real-World Problems — 131

6.2 Parabola: Equation and Applications — 132
- 6.2.1 Definition and Elements of the Parabola — 132
- 6.2.2 Equation of the Parabola in the Plane — 134
- 6.2.3 Applications of the Parabola in Physics and Geometry — 136

6.3 Ellipse and Hyperbola: Definition and Characteristics — **137**

- 6.3.1 Definition and Elements of the Ellipse 137
- 6.3.2 Definition and Elements of the Hyperbola 139
- 6.3.3 Applications of the Ellipse and Hyperbola 140

6.4 Solved Exercises — **141**

6.5 Proposed Exercises — **144**

- 6.5.1 Circle: Equation and Properties 144
- 6.5.2 Parabola: Equation and Applications 144
- 6.5.3 Ellipse and Hyperbola: Definitions and Characteristics 144

7 Functions: Domain, Range, and Operations 145

7.1 Domain and Range of Real Functions — **145**

- 7.1.1 Definition of Domain and Range 145
- 7.1.2 Determination of the Domain of Rational and Radical Functions 147
- 7.1.3 Graphical Representation of Domain and Range 148

7.2 Operations with Functions (Addition, Subtraction, Multiplication, Division) — **149**

- 7.2.1 Addition and Subtraction of Functions 149
- 7.2.2 Multiplication and Division of Functions 151
- 7.2.3 Properties of Operations with Functions 152

7.3 Piecewise-Defined Functions — **154**

- 7.3.1 Definition of Piecewise Functions 154
- 7.3.2 Graphical Representation of Piecewise-Defined Functions 156
- 7.3.3 Applications in Real-World Models 158

7.4 Solved Exercises — **159**

7.5 Proposed Exercises — **161**

- 7.5.1 Domain and Range of Real Functions 161
- 7.5.2 Operations with Functions (Sum, Subtraction, Multiplication, Division) 161
- 7.5.3 Piecewise-Defined Functions 161

8 Graphs and Graphical Transformations 163

8.1 Translation and Reflection of Functions — **163**

- 8.1.1 Vertical and Horizontal Translation 163
- 8.1.2 Reflection with Respect to the Axes 165
- 8.1.3 Graphical Representation of Transformations 167

8.2 Vertical and Horizontal Scaling — **169**

- 8.2.1 Scaling Functions Upward and Downward 169
- 8.2.2 Horizontal Scaling and Its Effects 172
- 8.2.3 Applications of Scaling in Graphs 174

8.3 Graphs of Composite Functions — **176**

- 8.3.1 Definition of Function Composition 176
- 8.3.2 Graph of Composite Functions 178
- 8.3.3 Examples of Composition and Its Applications 179

8.4 Solved Exercises — **180**

8.5 Proposed Exercises — **181**

- 8.5.1 Translation and Reflection of Functions 181
- 8.5.2 Vertical and Horizontal Scaling 181
- 8.5.3 Graphs of Composite Functions 181

9 Composition and Inverse Functions 183

9.1 Function Composition and Its Notation — 183
- 9.1.1 Definition of Function Composition 183
- 9.1.2 Notation and Properties of Composition 184
- 9.1.3 Examples of Function Composition 185

9.2 Inverse of a Function and Its Calculation — 186
- 9.2.1 Definition of an Inverse Function 186
- 9.2.2 Method for Finding the Inverse Function 188
- 9.2.3 Verification of the Inverse 190

9.3 Properties of Inverse Functions — 192
- 9.3.1 Relationship Between a Function and Its Inverse 192
- 9.3.2 Graph of an Inverse Function 195
- 9.3.3 Applications of Inverse Functions 195

9.4 Solved Exercises — 197

9.5 Proposed Exercises — 199
- 9.5.1 Function Composition and Notation 199
- 9.5.2 Inverse of a Function and Its Calculation 199
- 9.5.3 Properties of Inverse Functions 199

III Transcendental Functions, Matrices, and Systems of Equations

10 Transcendental Functions 203

10.1 Properties of Exponential and Logarithmic Functions — 203
- 10.1.1 Growth and Decay of Exponential Functions 203
- 10.1.2 Properties of Logarithms 205
- 10.1.3 Graphs of Exponential and Logarithmic Functions 206

10.2 Solving Exponential and Logarithmic Equations — 208
- 10.2.1 Basic Exponential Equations 208
- 10.2.2 Logarithmic Equations and Their Properties 209
- 10.2.3 Applications in Growth and Decay Problems 210

10.3 Trigonometric and Inverse Functions: Properties and Applications — 211
- 10.3.1 Definition of Trigonometric Functions 211
- 10.3.2 Fundamental Relationships Between Trigonometric Functions 213
- 10.3.3 Inverse Trigonometric Functions and Their Applications 215

10.4 Solved Exercises — 216

10.5 Proposed Exercises — 216
- 10.5.1 Properties of Exponential and Logarithmic Functions 216
- 10.5.2 Solving Exponential and Logarithmic Equations 217
- 10.5.3 Trigonometric and Inverse Functions: Properties and Applications 217

11 Matrices of Order 2 and 3 219

11.1 Matrix Operations (Addition, Subtraction, Multiplication) — 219
- 11.1.1 Matrix Addition and Subtraction 219
- 11.1.2 Matrix Multiplication and Properties 222
- 11.1.3 Transpose of a Matrix 224

11.2 Determinant Calculation and Adjoint of a Matrix — 226
- 11.2.1 Determinants of Matrices of Order 2 and 3 226
- 11.2.2 Determinant Properties 228
- 11.2.3 Adjoint of a Matrix 230

11.3 Inverse of a Matrix and Its Application in Systems of Equations — 232
- 11.3.1 Definition of the Inverse Matrix 232
- 11.3.2 Method for Calculating the Inverse (Gauss-Jordan) 234
- 11.3.3 Applications of the Inverse Matrix in Solving Systems 236

11.4 Ejercicios Resueltos — 238

11.5 Solved Exercises — 239

11.6 Proposed Exercises — 240
- 11.6.1 Matrix Operations (Addition, Subtraction, Multiplication) 240
- 11.6.2 Determinant and Adjoint Calculation of a Matrix 241
- 11.6.3 Inverse of a Matrix and Its Application in Systems of Equations 241

12 Systems of Linear and Nonlinear Equations 243

12.1 Methods for Solving Linear Systems — 243
- 12.1.1 Substitution Method and Examples 243
- 12.1.2 Step-by-Step Equalization Method 245
- 12.1.3 Elimination and System Reduction 247

12.2 Application of the Gauss-Jordan Method — 248
- 12.2.1 Solving Systems Using the Gauss-Jordan Method 248
- 12.2.2 Augmented Matrices and Row Reduction 250
- 12.2.3 Practical Examples of Solutions 253

12.3 Nonlinear Systems of Equations and Their Applications — 255
- 12.3.1 Definition of Nonlinear Systems 255
- 12.3.2 Methods for Solving Nonlinear Systems 257
- 12.3.3 Applications of Nonlinear Systems in Real-World Problems 258

12.4 Solved Exercises — 259

12.5 Proposed Exercises — 260
- 12.5.1 Methods for Solving Linear Systems 260
- 12.5.2 Application of Gauss-Jordan Method 260
- 12.5.3 Nonlinear Systems and Applications 261

Índice Alfabético 263

2. Summary

Basic Mathematics by Helbert Justo Luque Zevallos offers a comprehensive introduction to the essential principles of mathematics, structured in an accessible format for both students and professionals. This text covers a wide range of mathematical topics, from the fundamentals of algebra to advanced concepts in calculus and geometry,

Chapters:
- **Chapter 1: Real Number System** - Studies the basic properties and fundamental operations of real numbers.
- **Chapter 2: Equations and Inequalities** - Introduces methods for solving various forms of equations and inequalities.
- **Chapter 3: Complex Number System** - Explains the structure and operations within the set of complex numbers.
- **Chapter 4: Vectors in R2 and R3** - Addresses the theory and application of vectors in two and three dimensions.
- **Chapter 5: Analytical Geometry and Functions** - Develops the ability to describe and solve geometric and functional problems.
- **Chapter 6: Conic Equations** - Discusses the characteristics and equations of conics.
- **Chapter 7: Functions: Domain, Range, Operations** - Analyzes mathematical functions and their operations in detail.
- **Chapter 8: Graphs and Graphical Transformations** - Explores transformations and their effects on the graphs of functions.
- **Chapter 9: Composition and Inverse Functions** - Focuses on the composition of functions and determining inverse functions.
- **Chapter 10: Transcendental Functions, Matrices, and Systems of Equations** - Combines the study of transcendental functions with matrix theory and systems of equations.
- **Chapter 11: Matrices of Order 2 and 3** - Covers basic and advanced operations with small matrices.
- **Chapter 12: Systems of Linear and Nonlinear Equations** - Concludes with techniques for solving complex systems of equations.

3. Introduction

At the core of the exact and applied sciences lies a solid foundation in mathematics. *Basic Mathematics* is designed to provide undergraduate students in mathematics with a deep and applied understanding of the essential principles fundamental to various scientific and technological disciplines.

Methodology and Approach:
The approach adopted in this book is progressive and instructional. It begins with basic concepts and gradually advances to more complex topics. Each chapter builds on the knowledge of the previous one, enabling a gradual and complete understanding of each topic. Concepts are explained clearly, followed by detailed examples demonstrating their practical application. The exercises at the end of each chapter are designed to reinforce understanding and stimulate critical analysis.

Chapter Overview:
The book "Basic Mathematicsïs structured into twelve chapters designed to provide a comprehensive knowledge of key mathematical topics, with practical and theoretical applications:

1. **Real Numbers:** This chapter introduces the basic properties of real numbers, their operations, intervals, inequalities, and the concept of absolute value. The goal is to establish a solid foundation for understanding advanced mathematics.
2. **Equations and Inequalities:** Explores how to solve different types of linear, quadratic, and polynomial equations, as well as inequalities, including those with absolute values. The chapter aims to develop skills for manipulating and solving equations and inequalities that are fundamental in any field of mathematics.
3. **Complex Numbers:** Introduces the system of complex numbers, including their representation, basic operations, and properties. Students will learn to perform calculations in the complex plane, crucial knowledge for fields such as electrical engineering and physics.
4. **Vectors in R2 and R3:** Covers the fundamentals of vectors, their operations such as addition, scalar, and vector products, and their applications in geometric and physical contexts. The chapter aims for students to understand how vectors can model real-world situations and problems in multiple dimensions.
5. **Analytical Geometry and Functions:** Explains how to describe and analyze lines, planes,

and curves through equations. It focuses on analytical geometry techniques to solve problems of spatial location and geometric properties.

6. **Conic Equations:** Details the equations and properties of conics: circles, ellipses, parabolas, and hyperbolas. The chapter aims for students to identify and apply the characteristics of these curves in practical and theoretical problems.

7. **Functions: Domain, Range, Operations:** This chapter extensively covers the study of functions, including determining domains and ranges, operations between functions, and the analysis of composite and inverse functions. Through this, it aims to equip students with tools to analyze and construct functions in a mathematical context.

8. **Graphs and Graphical Transformations:** Delves into graph transformations such as translations, reflections, and dilations. Students will learn to interpret and manipulate function graphs to better understand their behavior.

9. **Composition and Inverse Functions:** Explores the composition of functions and inverse functions, detailing how these operations affect functions and their applications. This chapter helps students understand the relationship between functions and how to reverse them to find original solutions.

10. **Transcendental Functions, Matrices, and Systems of Equations:** Combines the study of transcendental functions such as exponential and logarithmic functions with an introduction to matrices and their use in solving systems of equations. The goal is to show how these mathematical tools are essential in modeling and solving complex systems.

11. **Matrices of Order 2 and 3:** Focuses on the study of small matrices, their manipulation, and how they are applied in contexts such as linear systems of equations. Students will learn methods to calculate determinants and inverses, key tools in solving linear systems.

12. **Systems of Linear and Nonlinear Equations:** The final chapter addresses the resolution of more complex systems using advanced techniques. It aims for students to apply numerical and algebraic methods to solve systems and analyze their consistency and possible solutions.

Each section is designed not only to teach mathematics but to demonstrate how these techniques are applicable in real-world situations, preparing students not only for academic exams but for professional challenges. When approaching this book, students are encouraged to read carefully, not skip worked examples, and extensively use the proposed problems for class discussion or individual study.

MBA. Helbert Justo Luque Zevallos

Mathematical Fundamentals and Algebra

1 Real Number System 17
1.1 Properties of Real Numbers
1.2 Intervals and Absolute Value
1.3 Operations with Real Numbers
1.4 Solved Exercises
1.5 Proposed Exercises

2 Equations and Inequalities 37
2.1 Solving Polynomial Equations
2.2 Equations with Radicals and Rationalization
2.3 Inequalities with Absolute Value and Their Graphical Representation
2.4 Solved Exercises
2.5 Proposed Exercises

3 Complex Number System 57
3.1 Representation of Complex Numbers in the Plane
3.2 Operations with Complex Numbers
3.3 De Moivre's Formula and Its Applications
3.4 Solved Exercises
3.5 Proposed Exercises

4 Vectors in $\mathbb{R}^2$ and $\mathbb{R}^3$ 83
4.1 Definition and Operations with Vectors
4.2 Dot Product and Cross Product
4.3 Parametric Equations and Cartesian Equations of Lines and Planes
4.4 Solved Exercises
4.5 Proposed Exercises

1. Real Number System

1.1 Properties of Real Numbers

1.1.1 Closure under Addition and Multiplication

Definition 1.1.1 The **closure** property states that for any operation performed between two elements of a set, the result always belongs to the same set. In the case of real numbers, the set is closed under addition and multiplication, meaning that the sum or product of two real numbers is always a real number.

This property is fundamental to defining operations within specific sets and ensuring the consistency of such operations. Closure extends not only to addition and multiplication but also to their combination with other algebraic properties.

■ **Example 1.1** Consider two real numbers, $a = 7$ and $b = -3$. The sum of these numbers, $a+b = 4$, belongs to the set of real numbers. Similarly, the product $a \cdot b = -21$ also belongs to the set of real numbers, confirming the closure property. ■

Lema 1.1.1 Let $a, b \in \mathbb{R}$. Then $a + b \in \mathbb{R}$ and $a \cdot b \in \mathbb{R}$. This is the **closure property** of real numbers, ensuring that the results of basic operations remain within the set of real numbers.

To understand the utility of the closure property, it is important to note that this property allows us to perform operations with confidence, knowing that we will always remain within the set of real numbers.

Theorem 1.1.1 For any pair of real numbers $a, b \in \mathbb{R}$, the sum $a+b$ and the product $a \cdot b$ are unique and belong to $\mathbb{R}$.

Demostración. To prove this theorem, we will demonstrate both the uniqueness and the membership of the sum and product of two real numbers in the set of real numbers.

Consider two real numbers $a, b \in \mathbb{R}$. The operation of addition, denoted by $a+b$, is defined for all real numbers. By the definition of real numbers and the closure property, we know that the sum of two real numbers is also a real number. This implies that $a+b \in \mathbb{R}$.

For uniqueness, note that the sum of two real numbers is well-defined and does not depend on the order (due to the commutative property of addition). That is, if we perform the operation $a+b$, we obtain a unique value in $\mathbb{R}$, as there are no two different results for the same operation.

Now consider the product of the same real numbers a and b. The operation of multiplication, denoted by $a \cdot b$, is also defined for all real numbers. The closure property ensures that the product of two real numbers belongs to $\mathbb{R}$. Therefore, $a \cdot b \in \mathbb{R}$.

For uniqueness, the commutative property of multiplication guarantees that $a \cdot b = b \cdot a$, and the result of the product of a and b is always a unique value. There are no two different results for the same multiplication operation between two real numbers.

Since both the sum and the product of two real numbers belong to $\mathbb{R}$ and their results are unique, we can conclude that the sum $a+b$ and the product $a \cdot b$ are unique and belong to the set of real numbers.

$$\therefore a+b \in \mathbb{R} \quad \text{and} \quad a \cdot b \in \mathbb{R} \quad \text{for any } a,b \in \mathbb{R}.$$

This completes the proof of the theorem. ∎

The above theorem reinforces the idea that operations within the set of real numbers are consistent and do not produce results outside the set. This has profound implications in measure theory, where we work with sets and functions defined on real numbers.

Corollary 1.1.2 Given that the set of real numbers is closed under addition and multiplication, any finite combination of these operations will also result in a real number.

This corollary is useful for understanding that the closure properties are preserved even when performing multiple consecutive operations. This facilitates the study of limits and continuity in more advanced contexts, as we will see in topics like integration.

Exercise 1.1 Prove that the set of rational numbers is closed under addition and multiplication. Provide examples illustrating this property.

Demostración. To prove that the set of rational numbers $\mathbb{Q}$ is closed under addition and multiplication, we must show that if $a,b \in \mathbb{Q}$, then $a+b \in \mathbb{Q}$ and $a \cdot b \in \mathbb{Q}$.

Let $a = \frac{p}{q}$ and $b = \frac{r}{s}$ be two rational numbers, where $p,q,r,s \in \mathbb{Z}$ and $q,s \neq 0$. The sum of a and b is calculated as follows:

$$a+b = \frac{p}{q} + \frac{r}{s} = \frac{ps+rq}{qs}$$

Since p,q,r,s are integers, so are ps and rq, and the product qs is non-zero. As $ps+rq$ and qs are integers with $qs \neq 0$, this implies that $a+b$ is a rational number, as it can be expressed as the quotient of two integers with a non-zero denominator. Hence, $a+b \in \mathbb{Q}$.

Similarly, the multiplication of a and b is:

$$a \cdot b = \frac{p}{q} \cdot \frac{r}{s} = \frac{p \cdot r}{q \cdot s}$$

Since $p,q,r,s \in \mathbb{Z}$ and $q,s \neq 0$, the product $p \cdot r$ and $q \cdot s$ are also integers, and $q \cdot s \neq 0$. This implies that $a \cdot b$ is a rational number, as it can be expressed as the quotient of two integers with a non-zero denominator. Hence, $a \cdot b \in \mathbb{Q}$.

Examples:

1.1 Properties of Real Numbers

1. Let $a = \frac{2}{3}$ and $b = \frac{3}{4}$. The sum is:

$$a + b = \frac{2}{3} + \frac{3}{4} = \frac{8+9}{12} = \frac{17}{12}$$

The result $\frac{17}{12}$ is a rational number.

2. Using the same numbers $a = \frac{2}{3}$ and $b = \frac{3}{4}$, the product is:

$$a \cdot b = \frac{2}{3} \cdot \frac{3}{4} = \frac{6}{12} = \frac{1}{2}$$

The result $\frac{1}{2}$ is a rational number.

Since the sum and product of two rational numbers always result in a rational number, we conclude that the set of rational numbers is closed under addition and multiplication. ∎

The link between the closure property and measure theory is evident, as a solid understanding of these basic properties provides a foundation for more advanced concepts, such as the properties of measurable functions and the construction of measures.

1.1.2 Distributive Property

Definition 1.1.2 The **distributive property** is one of the fundamental properties of arithmetic and algebra. It describes how multiplication interacts with addition or subtraction within an expression. For any real numbers a, b, and c, it holds that $a(b+c) = ab + ac$ and $a(b-c) = ab - ac$.

The distributive property is essential for simplifying algebraic expressions and solving equations. It directly connects with other properties such as associativity and commutativity, which will be studied later.

■ **Example 1.2** Consider the numbers $a = 3$, $b = 4$, and $c = 5$. Using the distributive property, we have:

$$3 \cdot (4+5) = 3 \cdot 4 + 3 \cdot 5 = 12 + 15 = 27$$

This shows how the distributive property allows us to break down an operation and compute it more easily. ∎

Lema 1.1.2 For any $a, b, c \in \mathbb{R}$, the distributive property implies that $a(b+c) = ab + ac$ and $a(b-c) = ab - ac$. This property holds for any numerical set closed under multiplication and addition.

The lemma above shows that the distributive property is not only valid for real numbers but also for other numerical systems, provided closure under the operations is maintained. This is particularly useful when working with polynomials and matrices.

Theorem 1.1.3 Let R be a commutative ring with unity. For any $a, b, c \in R$, the distributive property holds as $a(b+c) = ab + ac$.

Demostración. To prove the distributive property in a commutative ring R, take three arbitrary elements $a, b, c \in R$. The distributive property states that:

$$a(b+c) = ab + ac$$

In a ring, the addition operation $b+c$ is well-defined and belongs to the ring R. Since $b, c \in R$, it follows that $b+c \in R$ because R is closed under addition.

The multiplication $a(b+c)$ refers to multiplying element a with the result of the addition $b+c$. In a ring, multiplication is defined, and $a(b+c)$ belongs to R.

Applying the distributive property gives:

$$a(b+c) = a \cdot b + a \cdot c = ab + ac$$

By the definition of operations in the ring, we have multiplied a by each term of the sum $(b+c)$. In the first term, we multiply a by b, and in the second term, we multiply a by c. Since R is a commutative ring, the order of multiplication does not affect the result, and $ab, ac \in R$.

The result $ab + ac$ belongs to the ring R, and therefore, the distributive property holds in R. This implies that, for any pair of elements in a commutative ring with unity, multiplication distributes over addition:

$$a(b+c) = ab + ac$$

This completes the proof of the theorem. ∎

This theorem extends the distributive property to the context of ring theory, which is highly useful when studying more advanced algebraic structures. It allows us to apply this property to other sets such as polynomials or matrices.

Corollary 1.1.4 If $a, b, c \in \mathbb{R}$, then the distributive property also holds for successive multiplications, i.e., $a(bc + bd) = abc + abd$.

This corollary enables the application of the distributive property to more complex expressions, facilitating the expansion and simplification of terms in algebra.

> ® The distributive property is key for performing polynomial multiplication and factorization. Understanding this property will allow us to simplify calculations involving complex expressions, as well as develop efficient methods for solving equations.

Exercise 1.2 Use the distributive property to expand and simplify the following expression: $2(x+3) - 4(y-2)$. Then verify the result by calculating each term separately.

The concept of the distributive property connects to other important concepts, such as factorization and the expansion of algebraic expressions. In later topics, we will explore how to use this property in solving systems of equations and studying polynomial functions. Additionally, it is fundamental when working with matrices, where distribution allows for efficient multiplications.

1.1.3 Commutative and Associative Properties

Definition 1.1.3 The **commutative property** refers to the ability to swap operands without changing the result of the operation. This means that for any $a, b \in \mathbb{R}$, it holds that $a + b = b + a$ and $a \cdot b = b \cdot a$. On the other hand, the **associative property** states that the way terms are grouped in an operation does not affect the result: for $a, b, c \in \mathbb{R}$, $(a+b) + c = a + (b+c)$ and $(a \cdot b) \cdot c = a \cdot (b \cdot c)$.

These properties are fundamental in simplifying expressions and executing algebraic operations. By combining them, we can rearrange and group terms in the most convenient way, making calculations more efficient.

1.1 Properties of Real Numbers

■ **Example 1.3** Consider $a = 2$, $b = 3$, and $c = 4$. Applying the commutative property of addition, we have:

$$2 + 3 = 3 + 2 = 5$$

Similarly, for multiplication:

$$2 \cdot 3 = 3 \cdot 2 = 6$$

For the associative property, consider $(2+3)+4$:

$$(2+3) + 4 = 5 + 4 = 9$$

and also:

$$2 + (3+4) = 2 + 7 = 9$$

Both properties allow us to reorganize terms without changing the result. ■

Lema 1.1.3 For any $a, b, c \in \mathbb{R}$, the commutative and associative properties hold for both addition and multiplication. That is, $a + b = b + a$, $(a+b) + c = a + (b+c)$, $a \cdot b = b \cdot a$, and $(a \cdot b) \cdot c = a \cdot (b \cdot c)$.

This lemma ensures that we can reorder and regroup terms without affecting the final result, which is crucial when solving long equations and simplifying expressions.

Theorem 1.1.5 The commutative and associative properties are valid for any commutative ring. For any $a, b, c \in R$, where R is a commutative ring, it holds that $a + b = b + a$ and $(a+b) + c = a + (b+c)$.

Demostración. To prove this theorem, we need to verify both the commutative and associative properties of addition in a commutative ring R.

Take any $a, b \in R$, where R is a commutative ring. The commutative property of addition states that:

$$a + b = b + a$$

By the definition of a ring, addition in the ring R must satisfy the commutative property. This means that for any pair of elements $a, b \in R$, the sum of a and b does not depend on the order in which they are added. In other words, swapping the order of the addends does not affect the result.
Since R is a commutative ring, by definition, addition must satisfy this property. Therefore, we have $a + b = b + a$ for any $a, b \in R$.
Now consider three elements $a, b, c \in R$. The associative property of addition states that:

$$(a+b) + c = a + (b+c)$$

This means that the way elements are grouped in the addition does not affect the result. In a ring, addition is defined to be associative, meaning we can group the terms in the sum in any way and the result will remain the same.
To verify this, note that $(a+b) + c$ and $a + (b+c)$ represent the sum of the same elements a, b, c in different grouping orders, and since the addition operation in the ring R is associative by definition, we have:

$$(a+b) + c = a + (b+c)$$

Since we have shown that addition in a commutative ring R is both commutative and associative, we conclude that these properties hold for any commutative ring. Thus, for any $a, b, c \in R$:

$$a + b = b + a \quad \text{and} \quad (a+b) + c = a + (b+c)$$

This completes the proof of the theorem. ∎

This theorem is particularly relevant in studying more advanced algebraic structures, such as rings and fields, where these properties are essential for defining operations consistently.

Corollary 1.1.6 For any $a,b,c \in \mathbb{R}$, the commutative and associative properties allow simplifying complex algebraic expressions by rearranging and regrouping terms. For example, the expression $a+b+c$ can be evaluated in any order without changing the result.

This corollary reminds us that when manipulating algebraic expressions, we have the freedom to choose the order and grouping that are most convenient for simplifying calculations.

R The commutative and associative properties apply not only to real numbers but also to other algebraic structures such as matrices and polynomials, provided the operations are well-defined in those structures. This versatility makes these properties especially useful in linear algebra and polynomial theory.

Exercise 1.3 Verify whether the subtraction operation satisfies the commutative property. Provide a concrete example to justify your answer. Then, verify whether matrix multiplication satisfies the associative property.

The commutative and associative properties are closely related to other algebraic properties, such as the distributive property and the existence of identity elements. These connections are fundamental for building a solid foundation in algebra, which will be explored further in topics like factorization and solving systems of equations. Additionally, their understanding allows us to tackle advanced concepts in abstract algebra and linear algebra with greater depth.

1.2 Intervals and Absolute Value

1.2.1 Definition and Graphical Representation of Intervals

Definition 1.2.1 An **interval** is a set of real numbers contained between two endpoints. Depending on whether the endpoints are included or not, the interval can be open, closed, or half-open. A closed interval between a and b is denoted by $[a,b]$, and it includes both endpoints. In contrast, an open interval is denoted by (a,b), and it excludes both a and b.

The graphical representation of intervals on the number line provides a clear way to visualize the boundaries of a set. This is especially useful for understanding the inclusion or exclusion of specific values in an interval.

■ **Example 1.4** Let us represent the interval $(1,5]$. This interval includes all real numbers between 1 and 5, excluding 1 and including 5. The graphical representation is shown below:

Figura 1.2.1: *Graphical representation of the interval $(1,5]$.*

This example clearly shows how 1 is excluded, while 5 is included in the interval. ■

Lema 1.2.1 For any $a,b \in \mathbb{R}$ with $a < b$, the closed interval $[a,b]$ includes all real values x such that $a \leq x \leq b$. This ensures that all points within the endpoints belong to the defined set.

This lemma is fundamental for understanding how intervals work and how they can be used to define specific subsets of the number line.

1.2 Intervals and Absolute Value

> **Theorem 1.2.1** The intersection of two overlapping closed intervals is also a closed interval. That is, if $[a,b]$ and $[c,d]$ are two overlapping closed intervals, then their intersection is $[\text{máx}(a,c), \text{mín}(b,d)]$.

Demostración. Let $[a,b]$ and $[c,d]$ be two closed intervals in $\mathbb{R}$, and suppose the two intervals overlap. This means there are values of x that belong to both intervals.

To determine the intersection of the intervals, we need to consider the endpoints of both intervals. The intersection will consist of all points belonging to both interval $[a,b]$ and interval $[c,d]$.

To find the lower limit of the intersection interval, we take the larger value between the initial points a and c. This is denoted as $\text{máx}(a,c)$. This ensures that the lower limit is contained within both intervals.

Similarly, to find the upper limit of the intersection interval, we take the smaller value between the endpoints b and d. This is denoted as $\text{mín}(b,d)$. This ensures that the upper limit is contained within both intervals.

Thus, the intersection of the two intervals $[a,b]$ and $[c,d]$ is the closed interval defined by $[\text{máx}(a,c), \text{mín}(b,d)]$. It is worth noting that if $a \leq d$ and $c \leq b$, then the intervals indeed overlap, which guarantees that the intersection is non-empty and well-defined.

Therefore, we have proven that the intersection of two overlapping closed intervals is also a closed interval, given by $[\text{máx}(a,c), \text{mín}(b,d)]$. ∎

This theorem is highly useful in function analysis when determining common domains or studying the behavior of functions in a shared range.

> **Corollary 1.2.2** If $a,b,c \in \mathbb{R}$ with $a < b < c$, then the open interval (a,c) contains the closed interval $[b,c]$. This means that any number within the closed interval $[b,c]$ also belongs to the open interval (a,c).

Demostración. Consider the intervals (a,c) and $[b,c]$, with $a < b < c$.

The open interval (a,c) includes all real numbers x such that $a < x < c$. On the other hand, the closed interval $[b,c]$ contains all real numbers y such that $b \leq y \leq c$.

Since $a < b$, any value y in the interval $[b,c]$ satisfies $a < y \leq c$, which implies that y is also contained in the open interval (a,c). Note that the endpoint b of the closed interval $[b,c]$ is included in (a,c), as $b > a$.

Therefore, all points of the closed interval $[b,c]$ are contained in the open interval (a,c). This proves that (a,c) contains the interval $[b,c]$. ∎

This corollary highlights how open and closed intervals relate to each other, which is important when working with concepts like continuity and differentiability.

> **R** Intervals are fundamental building blocks in the construction of sets and function theory. Understanding how to represent and manipulate intervals facilitates the study of more advanced topics, such as the continuity of functions and the definition of the integral.

> **Exercise 1.4** Graph the following intervals on the number line:
> 1. $(2,6)$
> 2. $[-3,1)$
> 3. $[4,+\infty)$
>
> Then, find the intersection of the intervals $[-3,1)$ and $(2,6)$. Graph this intersection.

The understanding and representation of intervals is a key skill for any mathematics student, especially when studying functions and limits. In later chapters, we will explore how intervals play a crucial role in defining the domain of functions and in calculating limits and derivatives. Furthermore, intervals are also used to define partitions that are essential in the construction of the Riemann integral.

1.2.2 Types of Intervals (Open, Closed, Half-Open)

Definition 1.2.2 An **interval** is a subset of the number line that contains all the numbers between two given points, called endpoints. Intervals are classified as:
- **Open interval** (a,b): Includes all numbers between a and b but excludes the endpoints a and b.
- **Closed interval** $[a,b]$: Includes all numbers between a and b, including the endpoints a and b.
- **Half-open interval** $(a,b]$ or $[a,b)$: Includes all numbers between a and b, including only one of the endpoints.

Classifying intervals helps us understand different ways to constrain subsets of the number line, especially when studying limits and continuity.

■ **Example 1.5** Consider the following intervals:
- $(2,5)$: Includes all numbers between 2 and 5, excluding 2 and 5.
- $[0,3]$: Includes all numbers between 0 and 3, including 0 and 3.
- $(-1,4]$: Includes all numbers between -1 and 4, excluding -1 but including 4.

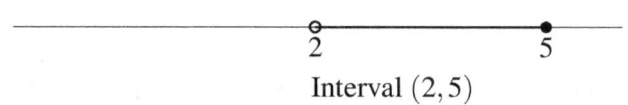

Interval $(2,5)$

Figura 1.2.2: *Graphical representation of the interval $(2,5)$.*

■

Lema 1.2.2 Every closed interval $[a,b]$ where $a < b$ always includes its endpoints, while open intervals (a,b) exclude both endpoints. Additionally, a half-open interval $(a,b]$ or $[a,b)$ includes only one of the endpoints. This property is fundamental when analyzing convergence on the number line.

The above lemma highlights the fundamental difference between the different types of intervals and their use in defining subsets. This difference will be key in mathematical analysis, where the inclusion or exclusion of points affects convergence and continuity.

Theorem 1.2.3 If $a,b \in \mathbb{R}$ with $a < b$, then any real number $x \in (a,b)$ also belongs to the closed interval $[a,b]$, except for the endpoints. In other words, $(a,b) \subseteq [a,b]$, and $[a,b]$ is the closure of (a,b).

Demostración. Consider the open interval (a,b) and the closed interval $[a,b]$, where $a < b$.
By definition, the open interval (a,b) includes all real numbers x such that $a < x < b$. Meanwhile, the closed interval $[a,b]$ includes all real numbers y such that $a \leq y \leq b$.
It is evident that any number x belonging to the open interval (a,b) also satisfies the condition to belong to the closed interval $[a,b]$, as x is between a and b, without reaching the endpoints. Therefore, we have $(a,b) \subseteq [a,b]$.
Additionally, the closed interval $[a,b]$ is obtained by adding the endpoints a and b to the open interval (a,b), which is why $[a,b]$ is referred to as the **closure** of (a,b).

1.2 Intervals and Absolute Value

Thus, it is concluded that all points of the open interval (a,b) are contained in the closed interval $[a,b]$, except for the endpoints, and that $[a,b]$ is the closure of (a,b). ∎

This theorem helps us understand the relationship between open and closed intervals. In particular, it shows that a closed interval contains all the points of its open counterpart plus the endpoints.

Corollary 1.2.4 For any open interval (a,b), the set of endpoints a and b does not belong to the interval. This means that the endpoints can always be added to form the closure of the interval.

This corollary reminds us of the importance of considering endpoints when working with open and closed sets, especially when studying functions with limits at the endpoints.

> The concept of open and closed intervals plays a crucial role in mathematical analysis. For example, when defining the continuity of a function, it is important to specify whether the endpoints of the interval in question are included. This is also applicable in integration, where the integration intervals must be carefully defined to ensure the existence and uniqueness of the integral.

Exercise 1.5 Graph the following intervals and determine whether they are open, closed, or half-open:
1. $[1,6)$
2. $(0,5]$
3. $(-\infty, 3]$

Then, analyze whether the intervals $[1,6)$ and $(0,5]$ have any points in common and, if so, determine which ones. ∎

Understanding the different types of intervals is essential to progressing toward more complex topics in mathematics. In the upcoming chapters, we will explore how open and closed intervals relate to the continuity of functions and the convergence of sequences. Additionally, in the context of measure theory, intervals serve as basic building blocks that allow us to construct more complex sets and define their measures.

1.2.3 Calculation and Properties of Absolute Value

Definition 1.2.3 The **absolute value** of a real number x, denoted by $|x|$, is defined as the distance of x from the origin on the number line. Mathematically, it is expressed as:

$$|x| = \begin{cases} x & \text{if } x \geq 0, \\ -x & \text{if } x < 0. \end{cases}$$

The absolute value is always a non-negative number, as it represents a distance.

This definition is fundamental for understanding how to measure distances on the number line, which is essential when working with inequalities and functions. Let us see some examples to illustrate the concept.

■ Example 1.6 Let us calculate the absolute value of the following numbers:
- $|3| = 3$ because 3 is positive.
- $|-5| = 5$ because the distance from -5 to the origin is 5.
- $|0| = 0$ because the origin has no distance from itself.

Graphically, the absolute value of a number can be represented as the distance from that number to the origin on the number line:

∎

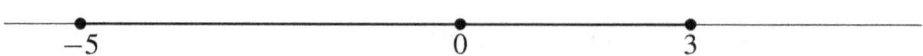

Figura 1.2.3: *Graphical representation of distances of absolute value.*

Lema 1.2.3 For any $a, b \in \mathbb{R}$, it holds that $|a \cdot b| = |a| \cdot |b|$. This means that the absolute value of a product is equal to the product of the absolute values.

This lemma is important because it allows us to manipulate algebraic expressions involving products within the absolute value, simplifying many algebraic operations and analyses.

> **Theorem 1.2.5** For any $a, b \in \mathbb{R}$, the triangle inequality holds:
>
> $$|a+b| \leq |a| + |b|$$
>
> The triangle inequality states that the total distance traveled when adding two real numbers is not greater than the sum of the individual distances.

Demostración. To prove the triangle inequality, we start by using the definition of absolute value. Recall that for any real number x:

$$|x| = \begin{cases} x & \text{if } x \geq 0, \\ -x & \text{if } x < 0. \end{cases}$$

Let $a, b \in \mathbb{R}$. First, note that by squaring the absolute value of the sum $(a+b)$, we have:

$$|a+b|^2 = (a+b)^2 = a^2 + 2ab + b^2.$$

On the other hand, consider the square of the sum of the absolute values $|a|$ and $|b|$:

$$(|a| + |b|)^2 = |a|^2 + 2|a||b| + |b|^2 = a^2 + 2|a||b| + b^2.$$

Since $|a| \geq 0$ and $|b| \geq 0$, it always holds that $|a||b| \geq ab$. Therefore, we have:

$$|a+b|^2 = a^2 + 2ab + b^2 \leq a^2 + 2|a||b| + b^2 = (|a| + |b|)^2.$$

Taking the square root on both sides of the inequality, we obtain:

$$|a+b| \leq |a| + |b|.$$

This concludes the proof of the triangle inequality for any $a, b \in \mathbb{R}$. ∎

This theorem is fundamental in many areas of mathematics, including analysis and geometry. It forms the basis of concepts such as sequence convergence and the calculation of norms in vector spaces.

> **Corollary 1.2.6** If $a, b \in \mathbb{R}$, then:
>
> $$||a| - |b|| \leq |a - b|$$

This corollary is a direct consequence of the triangle inequality and provides a useful tool for comparing differences between absolute values.

Demostración. To prove the corollary, we use the triangle inequality. Recall that for any $x, y \in \mathbb{R}$:

$$|x+y| \leq |x| + |y|.$$

Consider the expression $a - b$ and apply the triangle inequality:

$$|a| = |(a-b)+b| \leq |a-b| + |b|.$$

Subtracting $|b|$ on both sides of the inequality, we obtain:

$$|a| - |b| \leq |a-b|.$$

Similarly, consider the expression $b - a$:

$$|b| = |(b-a)+a| \leq |b-a| + |a| = |a-b| + |a|.$$

Subtracting $|a|$ on both sides, we obtain:

$$|b| - |a| \leq |a-b|.$$

Thus, we have two inequalities:

$$|a| - |b| \leq |a-b| \quad \text{and} \quad |b| - |a| \leq |a-b|.$$

We observe that this implies:

$$||a| - |b|| \leq |a-b|.$$

This concludes the proof of the corollary using the triangle inequality. ∎

> (R) The absolute value plays a crucial role when dealing with functions and limits, as it allows us to precisely define the distance between points on the number line. It is also useful in defining the norm in metric spaces, which is essential for studying topology and functional analysis.

> **Exercise 1.6** Calculate and graph the absolute values of the following expressions:
> 1. $|x-3|$ for $x = -2, 0, 4$.
> 2. $|2x+1|$ for $x = -1, 1, 3$.
>
> Then, verify the triangle inequality for $a = -2$ and $b = 4$.

Understanding the properties of absolute value is essential for tackling more complex topics, such as the analysis of piecewise-defined functions and solving inequalities. In later chapters, we will see how absolute value relates to the formal definition of limits and the continuity of functions, as well as its importance in the theory of integration and calculating distances in metric spaces.

1.3 Operations with Real Numbers

1.3.1 Addition and Subtraction of Real Numbers

> **Definition 1.3.1** **Addition** and **subtraction** of real numbers are fundamental operations that allow combining or comparing quantities. The addition of two real numbers a and b is defined as their result $a + b$, while the subtraction $a - b$ is defined as the addition of a with the opposite of b, that is, $a + (-b)$.

Understanding addition and subtraction is essential, as these operations form the basis for more complex algebraic operations such as multiplication, division, and exponentiation.

■ **Example 1.7** Consider the numbers $a = 4$ and $b = -3$. The addition of these two numbers is:

$$4 + (-3) = 1$$

While the subtraction is:

$$4 - (-3) = 4 + 3 = 7$$

These operations can be graphically represented on the number line:

Figura 1.3.1: *Graphical representation of addition and subtraction on the number line.*

■

Lema 1.3.1 Addition of real numbers is **commutative** and **associative**. That is, for any $a, b, c \in \mathbb{R}$, the following hold:

$$a + b = b + a$$

and

$$(a + b) + c = a + (b + c)$$

This lemma ensures that real numbers can be added in any order or grouping without affecting the result. This is crucial when dealing with multiple terms.

> Theorem 1.3.1 For any $a, b, c \in \mathbb{R}$, subtraction satisfies the following property:
>
> $$a - (b + c) = (a - b) - c$$
>
> This property shows how subtraction can be distributed over multiple terms.

Demostración. To prove the property, we evaluate both sides of the equation.
First, compute $a - (b + c)$:

$$a - (b + c) = a - b - c$$

Now compute $(a - b) - c$:

$$(a - b) - c = a - b - c$$

Since both sides are equal, we conclude:

$$a - (b + c) = (a - b) - c$$

This proves the property of the distribution of subtraction. ■

The above theorem simplifies expressions involving consecutive subtractions, allowing a structured approach to breaking down terms.

1.3 Operations with Real Numbers

Corollary 1.3.2 If $a, b \in \mathbb{R}$, then:

$$a - b = -(b - a)$$

This corollary indicates that subtracting two numbers is equivalent to the opposite of subtracting them in reverse order.

This property is useful when reorganizing terms in algebraic expressions and solving equations involving subtractions between different terms.

> (R) Addition and subtraction of real numbers are operations that have significant applications in calculating limits, derivatives, and analyzing functions. It is important to note that addition is commutative, whereas subtraction is not, which influences the resolution of equations and the simplification of expressions.

Exercise 1.7 Calculate and graphically represent the following results:
1. $7 + (-2) - 5$
2. $3 - (4 + 6)$
3. $-5 + (-3) - (-1)$

Verify the commutative property of addition for the first two terms of each expression.

A deep understanding of the properties of addition and subtraction of real numbers is essential for solving equations, analyzing functions, and evaluating limits. In subsequent chapters, we will explore how these operations generalize to larger sets, such as matrices and vectors, where the properties of addition play a crucial role in linear algebra and analytical geometry.

1.3.2 Multiplication and Division of Real Numbers

Definition 1.3.2 Multiplication of real numbers is an operation that, for two numbers a and b, produces a third number called the product, denoted by $a \cdot b$ or simply ab. **Division**, on the other hand, is the inverse operation of multiplication and is defined for two real numbers a and b, with $b \neq 0$, as a/b, which represents how many times b is contained in a.

Multiplication and division are fundamental in manipulating algebraic expressions and are widely used in solving equations and analyzing functions.

■ **Example 1.8** Consider the numbers $a = 6$ and $b = -3$. The multiplication of these two numbers is:

$$6 \cdot (-3) = -18$$

While the division is:

$$6/(-3) = -2$$

We can visualize multiplication on a number line by extending 6 in -3 steps to the left, resulting in -18. ■

Lema 1.3.2 Multiplication of real numbers is **commutative** and **associative**. That is, for any $a, b, c \in \mathbb{R}$, the following hold:

$$a \cdot b = b \cdot a$$

and

$$(a \cdot b) \cdot c = a \cdot (b \cdot c)$$

This lemma ensures that the multiplication of real numbers does not depend on the order or grouping, which facilitates the simplification of products and solving algebraic problems.

> **Theorem 1.3.3** For any $a, b, c \in \mathbb{R}$, with $b, c \neq 0$, division satisfies the following property:
>
> $$\frac{a}{b} \cdot \frac{b}{c} = \frac{a}{c}$$
>
> This theorem shows how fractions multiplied together can be simplified.

Demostración. Consider the expression:

$$\frac{a}{b} \cdot \frac{b}{c}$$

To prove this is equal to $\frac{a}{c}$, multiply the two fractions. Recall that to multiply fractions, we multiply the numerators and denominators respectively:

$$\frac{a}{b} \cdot \frac{b}{c} = \frac{a \cdot b}{b \cdot c}$$

In the numerator and denominator, there is a common factor b. Since $b \neq 0$, we can simplify:

$$\frac{a \cdot b}{b \cdot c} = \frac{a}{c}$$

Thus, we have shown that:

$$\frac{a}{b} \cdot \frac{b}{c} = \frac{a}{c}$$

Therefore, the property is satisfied.
∎

The above theorem is useful in calculating and manipulating rational expressions, providing a powerful tool for simplifying fractions and solving equations.

> **Corollary 1.3.4** If $a, b \in \mathbb{R}$ and $b \neq 0$, then:
>
> $$\frac{a}{b} = a \cdot \frac{1}{b}$$
>
> This corollary implies that the division of two real numbers can be expressed as the multiplication by the reciprocal of the divisor.

This corollary reminds us that division and multiplication are intimately related, and that the division operation can be understood as multiplication by an inverse.

> (R) Multiplication and division of real numbers play an important role in the development of algebra, as they are used in solving equations and factoring algebraic expressions. Additionally, they are essential in the study of rational functions, where understanding how to divide and multiply terms allows us to simplify and better analyze functions.

1.3 Operations with Real Numbers

> **Exercise 1.8** Calculate and graphically represent the following results:
> 1. $(-4) \cdot 3 + 2$
> 2. $12/(-3)$
> 3. $\frac{-6}{2} \cdot (-1)$
>
> Verify the commutative property of multiplication and the relationship between division and multiplication for the first two terms of each expression.

Understanding the properties of multiplication and division is key to advancing in the study of algebra and mathematical analysis. In subsequent chapters, we will apply these properties to the study of polynomials, algebraic fractions, and rational functions. Additionally, we will explore how multiplying terms affects the shape and behavior of functions in graphical analysis and in solving optimization problems.

1.3.3 Powers and Roots of Real Numbers

> **Definition 1.3.3** A **power** of a real number a, denoted as a^n, is the result of multiplying a by itself n times, where n is a positive integer. If $n = 0$, then $a^0 = 1$ (for $a \neq 0$). A **root** of a real number a, denoted as $\sqrt[n]{a}$, is a number b such that $b^n = a$. The square root ($\sqrt{a}$) and cube root ($\sqrt[3]{a}$) are common examples of roots.

Powers and roots are inverse operations and play an important role in solving algebraic equations and analyzing functions.

■ **Example 1.9** Consider the following calculations:
- The power 2^3 is calculated as $2 \times 2 \times 2 = 8$.
- The square root of 16 is $\sqrt{16} = 4$, since $4^2 = 16$.
- The cube root of 27 is $\sqrt[3]{27} = 3$, since $3^3 = 27$.

We can visually represent the relationship between a power and its root as a function on the number line:

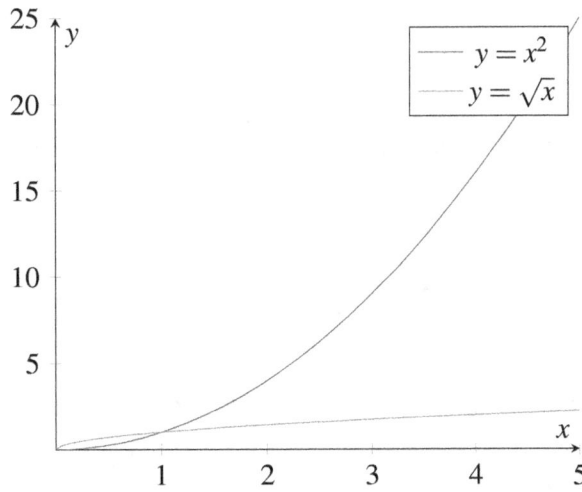

Figura 1.3.2: *Graphical representation of the functions $y = x^2$ and $y = \sqrt{x}$.*

■

Lema 1.3.3 For any real numbers $a, b \in \mathbb{R}$ and $n \in \mathbb{N}$, the following holds:

$$a^n \cdot b^n = (a \cdot b)^n$$

This property demonstrates how powers with the same base can be combined efficiently.

The above lemma is key to simplifying algebraic expressions and has applications in solving exponential equations and factorization.

> **Theorem 1.3.5** For any real number $a > 0$ and any integers $m, n \in \mathbb{Z}$ with $n \neq 0$, the following fractional power property holds:
>
> $$a^{m/n} = \sqrt[n]{a^m} = (\sqrt[n]{a})^m$$
>
> This property allows us to express powers with fractional exponents in terms of roots.

Demostración. Consider the property we want to prove:

$$a^{m/n} = \sqrt[n]{a^m} = (\sqrt[n]{a})^m$$

To prove this property, we can decompose the fractional exponent into two parts. Recall that $a^{m/n}$ represents a fractional exponent where the denominator indicates a root and the numerator indicates a power.

By definition, $a^{m/n}$ means raising a to the power of m and then taking the nth root of the result, which can be written as:

$$a^{m/n} = \sqrt[n]{a^m}$$

This means we first raise a to the power of m and then take the nth root of that value.

The property of roots also allows us to change the order of the exponents. That is, we can first take the nth root of a and then raise the result to the power of m:

$$\sqrt[n]{a^m} = (\sqrt[n]{a})^m$$

This equality is based on the definition of fractional exponents and the commutative property of powers and roots.

From the above, we have:

$$a^{m/n} = \sqrt[n]{a^m} = (\sqrt[n]{a})^m$$

Therefore, the property holds for any positive real number $a > 0$ and any integers m, n with $n \neq 0$. ∎

The theorem is particularly useful for simplifying powers that involve roots, which is common in calculus and mathematical analysis.

> **Corollary 1.3.6** If $a > 0$ and $n \in \mathbb{N}$, then:
>
> $$(\sqrt[n]{a})^n = a$$
>
> This corollary indicates that the power of a root of a number always returns the original value if the exponent matches the index of the root.

1.3 Operations with Real Numbers

Demostración. Consider the expression:

$$(\sqrt[n]{a})^n$$

By definition, $\sqrt[n]{a}$ represents the number that, when raised to the power of n, results in a. That is:

$$\sqrt[n]{a} = a^{1/n}$$

Now, raise $\sqrt[n]{a}$ to the power of n:

$$(\sqrt[n]{a})^n = (a^{1/n})^n$$

Using the property of exponents, $(a^m)^n = a^{m \cdot n}$, we can simplify the expression as follows:

$$(a^{1/n})^n = a^{(1/n) \cdot n} = a^1 = a$$

Thus, we have demonstrated that:

$$(\sqrt[n]{a})^n = a$$

This confirms that the power of a root of a number always returns the original value when the exponent matches the root index. ∎

This property is crucial when eliminating roots and solving algebraic equations involving powers and roots.

> **R** The properties of powers and roots are not only applicable to positive real numbers but also extend to other mathematical fields, such as complex numbers, where roots of negative numbers are defined. This extension has significant applications in equation theory and solving problems related to polynomial functions.

> **Exercise 1.9** Calculate and graphically represent the following results:
> 1. $3^2 - \sqrt{9}$
> 2. $\sqrt[3]{8} \cdot 2^3$
> 3. $(-2)^4$ and $\sqrt[4]{16}$
>
> Verify the fractional power property in the second exercise.

The study of powers and roots is fundamental to understanding the behavior of exponential and polynomial functions. In subsequent chapters, we will explore how these properties are essential for solving algebraic equations, analyzing complex functions, and developing advanced concepts in differential and integral calculus. Additionally, understanding these properties will enable us to better handle logarithmic and hyperbolic functions, which are of great importance in various areas of mathematics and their applications.

1.4 Solved Exercises

Exercise 1.10 Prove that the distributive property holds for all real numbers a, b, and c: $a(b+c) = ab+ac$.

Demostración. The distributive property states that for any real numbers a, b, and c, multiplication distributes over addition as follows:

$$a(b+c) = ab+ac$$

To prove this, expand the left-hand side of the equation:

$$a(b+c) = a \cdot b + a \cdot c = ab + ac$$

This demonstrates that the product of a with the sum $(b+c)$ is equivalent to the sum of the individual products ab and ac. This equality holds for all real numbers a, b, and c, completing the proof. ∎

Exercise 1.11 Solve the equation $|x-3| = 5$ and determine the values of x. Represent the solution on the number line.

Demostración. To solve $|x-3| = 5$, we consider two possibilities due to the definition of absolute value:

$$x-3 = 5 \quad \text{or} \quad x-3 = -5$$

Solving both equations:
- $x-3 = 5 \Rightarrow x = 8$
- $x-3 = -5 \Rightarrow x = -2$

Thus, the solutions are $x = 8$ and $x = -2$.
Representing these solutions on the number line, we have two points: $x = -2$ and $x = 8$. ∎

Exercise 1.12 Find the values of x that satisfy the inequality $\frac{x^2-4}{x-2} > 0$. Represent the solutions on the number line.

Demostración. First, factorize the numerator:

$$x^2 - 4 = (x-2)(x+2)$$

Thus, the inequality becomes:

$$\frac{(x-2)(x+2)}{x-2} > 0$$

To simplify the expression, cancel the factor $(x-2)$, but note that $x \neq 2$ as this would make the denominator zero. We are left with:

$$x + 2 > 0$$

Solving this inequality:

$$x > -2$$

Additionally, $x \neq 2$ due to the restriction on the denominator. Therefore, the solution is:

$$x > -2, \quad x \neq 2$$

Representing this on the number line, the solution interval is $(-2, 2) \cup (2, \infty)$, with a gap at $x = 2$. ∎

1.5 Proposed Exercises

Exercise 1.13 Consider two intervals $A = [1,4]$ and $B = (2,5)$. Find the intersection and union of these intervals.

Demostración. ■ **Intersection**:

The intersection of intervals $A = [1,4]$ and $B = (2,5)$ is the set of values that belong to both intervals. Therefore, the intersection is:

$$A \cap B = (2,4]$$

- **Union**:

The union of intervals A and B is the set of all values that belong to at least one of the two intervals. Therefore, the union is:

$$A \cup B = [1,5)$$

■

Exercise 1.14 Prove that for any real numbers a, b, if $a \leq b$, then $a + c \leq b + c$ for any $c \in \mathbb{R}$.

Demostración. Starting with the inequality $a \leq b$, add the real number c to both sides:

$$a + c \leq b + c$$

The order property of real numbers allows us to add the same quantity to both sides of an inequality without changing the inequality's direction. This property holds for any real number c, completing the proof. ■

1.5 Proposed Exercises

1.5.1 Properties of Real Numbers

Exercise 1.15 Prove that for any real numbers a, b, and c, the distributive property holds: $a(b+c) = ab + ac$.

Exercise 1.16 Prove that the addition operation is commutative in the set of real numbers, i.e., $a + b = b + a$ for any $a, b \in \mathbb{R}$.

Exercise 1.17 Determine whether multiplication in the set of real numbers satisfies the associative property. Justify your answer with examples.

Exercise 1.18 Prove that if a and b are real numbers and $a \leq b$, then $a + c \leq b + c$ for any real number c.

Exercise 1.19 Find a counterexample to show that the commutative property does not hold for subtraction in the set of real numbers.

1.5.2 Intervals and Absolute Value

Exercise 1.20 Graph the intervals $(-\infty, 3]$ and $[1, 5)$ on the number line and determine their intersection.

Exercise 1.21 Prove that $|a-b| \geq ||a|-|b||$ for any real numbers a and b.

Exercise 1.22 Determine all values of x that satisfy the inequality $|x-3| < 5$.

Exercise 1.23 Solve the equation $|2x+1| = 7$ and represent the solution on the number line.

Exercise 1.24 Describe geometrically what $|x-4| < 3$ represents on the number line.

1.5.3 Operations with Real Numbers

Exercise 1.25 Find the values of a and b such that $(a+b)^2 = 25$ and $a-b = 3$.

Exercise 1.26 Simplify and solve the expression $(2x+3)^2 - (x-1)^2$.

Exercise 1.27 Prove that $x^2 + y^2 \geq 0$ for any $x, y \in \mathbb{R}$.

Exercise 1.28 Simplify $\frac{(x^2-4)(x+2)}{x-2}$.

Exercise 1.29 Find all possible values of x such that $x^2 - 5x + 6 = 0$.

2. Equations and Inequalities

2.1 Solving Polynomial Equations

2.1.1 First-Degree Equations

Definition 2.1.1 A **first-degree equation** is an equation that can be expressed in the general form $ax + b = 0$, where a and b are real numbers and $a \neq 0$. The variable x is a real number that satisfies the equation. First-degree equations are also known as *linear equations* because their graph is a straight line in the Cartesian coordinate system.

First-degree equations are fundamental in the study of basic algebra, and solving them involves finding the value of the variable that satisfies the equality.

■ **Example 2.1** Consider the first-degree equation $3x - 6 = 0$. To solve it, we isolate x:

$$3x = 6$$

$$x = \frac{6}{3} = 2$$

The value $x = 2$ is the solution to the equation, which can be verified by substituting $x = 2$ into the original equation and confirming that the equality holds. ■

Lema 2.1.1 For any first-degree equation $ax + b = 0$, where $a, b \in \mathbb{R}$ and $a \neq 0$, the solution is given by:

$$x = -\frac{b}{a}$$

This lemma provides a general method for solving any first-degree equation by isolating x. This technique is useful for solving linear equations quickly without extensive manipulation.

Theorem 2.1.1 If $ax + b = 0$ and $cx + d = 0$ are two first-degree equations with $a, b, c, d \in \mathbb{R}$, then these equations have the same solution if and only if:

$$a = c \quad \text{and} \quad b = d$$

Demostración. Assume that the equations $ax+b=0$ and $cx+d=0$ have the same solution. Denote this solution by x_0. Then:

$$ax_0 + b = 0 \quad \text{and} \quad cx_0 + d = 0$$

Since x_0 is the same solution for both equations, we can equate their expressions:

$$ax_0 + b = cx_0 + d$$

Subtracting ax_0 and cx_0 from both sides, we get:

$$b = d$$

And subtracting b from both sides, we obtain:

$$a = c$$

Therefore, $a = c$ and $b = d$.
Conversely, if $a = c$ and $b = d$, then the equations are identical:

$$ax + b = cx + d$$

Thus, they have the same solution for any value of x. ∎

This theorem establishes when two first-degree equations can be considered equivalent, which is important for understanding algebraic transformations that do not alter the solution of the equation.

Corollary 2.1.2 If $a \neq 0$, then the equation $ax = 0$ has a unique solution given by $x = 0$. This implies that if the constant term is zero, the only possible solution to the equation is zero.

This corollary is useful when dealing with homogeneous equations, where the constant term is zero. This property is particularly relevant in linear systems and linear algebra.

 First-degree equations are the foundation for studying more advanced topics such as quadratic equations and systems of linear equations. Additionally, solving first-degree equations is a key skill for tackling optimization problems, physics equations, and economic models.

Exercise 2.1 Solve the following first-degree equations:
1. $5x + 7 = 0$
2. $-3x + 9 = 12$
3. $\frac{1}{2}x - 4 = 0$

Then, verify the solutions graphically by representing each equation on the number line.

The study of first-degree equations is essential for advancing in algebra and understanding the behavior of linear functions. In the next chapters, we will explore how these equations apply to solving systems of equations and how they can be used to model real-world situations, such as rectilinear motion problems or financial scenarios solvable with linear models.

2.1.2 Second-Degree Equations

Definition 2.1.2 A **second-degree equation** is an equation that can be written in the general form $ax^2 + bx + c = 0$, where $a, b, c \in \mathbb{R}$ and $a \neq 0$. The variable x is the value that satisfies the equality. Second-degree equations are also called *quadratic equations*, and their graphical representation is a parabola.

There are different methods to solve second-degree equations, with the most common being factoring and using the quadratic formula. Both methods complement each other and provide efficient ways to solve these equations.

■ **Example 2.2** Consider the equation $x^2 - 5x + 6 = 0$. We can solve it by factoring:

$$x^2 - 5x + 6 = (x-2)(x-3) = 0$$

The solutions are $x = 2$ and $x = 3$. We can easily verify that both roots satisfy the original equation.

■

Lema 2.1.2 The quadratic equation $ax^2 + bx + c = 0$ has two real roots if and only if the discriminant $b^2 - 4ac \geq 0$. If $b^2 - 4ac = 0$, the equation has a double root, while if $b^2 - 4ac > 0$, it has two distinct roots.

The discriminant is a key tool to determine the nature of the roots of a second-degree equation. It indicates whether the roots are real and distinct, real and equal, or complex.

Theorem 2.1.3 The **quadratic formula** for solving a second-degree equation $ax^2 + bx + c = 0$ is:

$$x = \frac{-b \pm \sqrt{b^2 - 4ac}}{2a}$$

This formula provides a systematic way to find the roots of any quadratic equation, provided that $a \neq 0$.

Demostración. To prove the quadratic formula for the equation $ax^2 + bx + c = 0$, we use the method of completing the square.

First, divide the entire equation by a (assuming $a \neq 0$):

$$x^2 + \frac{b}{a}x + \frac{c}{a} = 0$$

Move the constant term to the other side of the equation:

$$x^2 + \frac{b}{a}x = -\frac{c}{a}$$

To complete the square, add and subtract the appropriate term. The term to add is $\left(\frac{b}{2a}\right)^2$:

$$x^2 + \frac{b}{a}x + \left(\frac{b}{2a}\right)^2 = -\frac{c}{a} + \left(\frac{b}{2a}\right)^2$$

This allows us to rewrite the left-hand side as a perfect square:

$$\left(x + \frac{b}{2a}\right)^2 = \frac{b^2 - 4ac}{4a^2}$$

Take the square root of both sides:

$$x + \frac{b}{2a} = \pm \frac{\sqrt{b^2 - 4ac}}{2a}$$

Finally, solve for x:

$$x = \frac{-b \pm \sqrt{b^2 - 4ac}}{2a}$$

This is the quadratic formula for solving a second-degree equation. ∎

This theorem is fundamental for solving any second-degree equation, even when factoring is not straightforward. The formula guarantees a solution for all values of $a, b, c \in \mathbb{R}$.

Corollary 2.1.4 If $b^2 - 4ac < 0$, the roots of the quadratic equation $ax^2 + bx + c = 0$ are complex conjugates, given by:

$$x = \frac{-b \pm i\sqrt{4ac - b^2}}{2a}$$

This implies that quadratic equations with a negative discriminant have no real solutions.

Complex roots have significant applications in algebra and analysis, especially when solving differential equations or studying the behavior of functions in the complex plane.

> (R) Second-degree equations appear in various contexts, from problems involving parabolic motion in physics to optimization in economics. Understanding how to solve them using factoring and the quadratic formula is essential for tackling more complex problems in algebra and calculus.

> **Exercise 2.2** Solve the following second-degree equations using the quadratic formula and factoring where possible:
> 1. $x^2 - 7x + 10 = 0$
> 2. $2x^2 + 4x - 6 = 0$
> 3. $x^2 + 4x + 5 = 0$
>
> Then, verify the nature of the roots using the discriminant.

Second-degree equations form a fundamental step in transitioning to higher-degree equations and nonlinear systems. In later chapters, we will explore how these equations apply to analyzing quadratic functions and how they relate to analytic geometry, particularly in determining points of intersection between curves.

2.1.3 Higher-Degree Equations

Definition 2.1.3 A **higher-degree equation** is an equation in the general form $a_n x^n + a_{n-1} x^{n-1} + \cdots + a_1 x + a_0 = 0$, where $a_i \in \mathbb{R}$ and $a_n \neq 0$, with $n > 2$. The degree of the equation is determined by the highest exponent of the variable x. Such equations are typically solved through factoring, substitution, or more advanced techniques like the use of complex roots.

Higher-degree equations are a natural extension of quadratic equations and present a greater challenge, as they can have multiple real or complex roots.

2.1 Solving Polynomial Equations

■ **Example 2.3** Consider the cubic equation $x^3 - 6x^2 + 11x - 6 = 0$. We can solve it by factoring:

$$x^3 - 6x^2 + 11x - 6 = (x-1)(x-2)(x-3) = 0$$

From this, the solutions are $x = 1$, $x = 2$, and $x = 3$. These roots indicate the points where the function intersects the x-axis. ■

Lema 2.1.3 If $f(x)$ is a polynomial of degree $n > 1$ and has a root r, then $f(x)$ can be factored as $(x-r)g(x)$, where $g(x)$ is a polynomial of degree $n-1$. This is known as the **Factor Theorem**.

Demostración. Given that $f(x)$ is a polynomial of degree $n > 1$ and has a root r, by definition, $f(r) = 0$.
According to the Factor Theorem, if r is a root of $f(x)$, then there exists a polynomial $g(x)$ of degree $n-1$ such that $f(x)$ can be written as:

$$f(x) = (x-r)g(x)$$

The existence of $g(x)$ is due to the exact divisibility of $f(x)$ by $(x-r)$, since $f(r) = 0$. The polynomial $g(x)$ has degree $n-1$ because dividing a degree n polynomial by a linear factor reduces its degree by 1. ∎

This lemma is extremely useful for breaking down higher-degree equations into simpler factors, making it easier to find roots and reduce the degree of the equation.

Theorem 2.1.5 The **Fundamental Theorem of Algebra** states that any polynomial of degree n with real coefficients has exactly n roots, including real and complex roots, and accounting for multiplicity. This implies that every higher-degree equation has at least one solution in the set of complex numbers.

The Fundamental Theorem of Algebra cannot be fully proven through elementary methods, as its proof involves advanced concepts in complex analysis. However, a simplified explanation outlines the idea behind the theorem:

Demostración. Suppose we have a polynomial $p(x)$ of degree n with real or complex coefficients. We aim to prove that there is at least one complex root for this polynomial. One way to prove this is by considering the behavior of $|p(x)|$, which grows large as $|x|$ increases significantly. This implies that $p(x)$ reaches a minimum value at some point in the complex plane.
If this minimum value were nonzero, the polynomial would have no roots, contradicting the theorem. Therefore, there must exist at least one point where $p(x) = 0$. Using the Factor Theorem (which states that if r is a root, the polynomial can be factored by $(x-r)$), we can reduce the degree of the polynomial. Repeating this process n times proves that the polynomial has exactly n roots, counting multiplicities. ∎

This theorem guarantees that we can always find all roots of a polynomial, which is fundamental for solving higher-degree equations and understanding their behavior.

Corollary 2.1.6 If $f(x)$ is a polynomial of degree n with real coefficients, then the complex roots of $f(x)$ occur in conjugate pairs. That is, if $a + bi$ is a root, then $a - bi$ is also a root.

Demostración. Let $f(x)$ be a polynomial of degree n with real coefficients, and suppose $z_1 = a + bi$ with $a, b \in \mathbb{R}$ and $b \neq 0$ is a root of $f(x)$, meaning:

$$f(z_1) = 0$$

Since the coefficients of $f(x)$ are real, taking the conjugate of the expression $f(z_1)$ gives:

$$f(\overline{z_1}) = \overline{f(z_1)} = \overline{0} = 0$$

This implies that $\overline{z_1} = a - bi$ is also a root of $f(x)$.
Thus, if $f(x)$ has real coefficients and a complex root $a + bi$, its conjugate $a - bi$ is also a root. ∎

This corollary is particularly important when working with higher-degree polynomials with real coefficients, as it determines the structure of the complex roots.

> (R) Higher-degree equations have numerous applications, from modeling physical phenomena to solving financial problems. Understanding how to find their roots and factorize polynomials is essential for tackling more advanced problems, such as optimizing functions and studying the stability of dynamic systems.

> **Exercise 2.3** Solve the following higher-degree equations using factoring and the Factor Theorem:
> 1. $x^3 - 4x^2 - 7x + 10 = 0$
> 2. $x^4 - 5x^3 + 6x^2 + 4x - 8 = 0$
> 3. $2x^3 + x^2 - 8x - 4 = 0$
>
> Then, verify the nature of the roots using the Fundamental Theorem of Algebra and discuss whether there are any complex roots.

Understanding higher-degree equations and their solution methods is crucial for advancing in algebra and analysis. In the upcoming chapters, we will explore how these equations apply to curve analysis, polynomial function theory, and numerical methods for finding approximate solutions. Furthermore, we will see how these equations are fundamental in physics for modeling the behavior of nonlinear systems and solving dynamic problems.

2.2 Equations with Radicals and Rationalization

2.2.1 Definition of Radicals

> **Definition 2.2.1** A **radical** is an expression involving a root, such as a square root, cube root, or root of order n. The root of a number a, denoted as $\sqrt[n]{a}$, is defined as the value which, raised to the power n, equals the number a. Specifically, the **square root** of a is denoted by $\sqrt{a}$ and represents the non-negative value x such that $x^2 = a$.

Radicals are fundamental in algebra and frequently appear in the simplification of expressions and the solving of equations. Let us see some examples to better illustrate this concept.

■ **Example 2.4** Consider the following radicals:
- $\sqrt{16} = 4$, because $4^2 = 16$.
- $\sqrt[3]{27} = 3$, because $3^3 = 27$.
- $\sqrt[4]{81} = 3$, because $3^4 = 81$.

These examples demonstrate how different roots are used to find values that, when raised to a specific power, yield the original number. ■

Lema 2.2.1 For any non-negative real numbers $a, b \in \mathbb{R}$ and a positive integer n, the following property holds:

$$\sqrt[n]{a \cdot b} = \sqrt[n]{a} \cdot \sqrt[n]{b}$$

This property allows us to simplify products within radicals into products of individual radicals.

2.2 Equations with Radicals and Rationalization

This property is useful for simplifying complex expressions involving products of radicals and is the foundation for many developments in algebra and analysis.

> **Theorem 2.2.1** The operation of extracting roots satisfies the property of fractional exponents. That is, for any positive real number $a > 0$ and a positive integer n, the following holds:
> $$\sqrt[n]{a} = a^{1/n}$$
> This property establishes a direct relationship between radicals and powers, allowing the simplification of expressions involving roots using fractional exponents.

Demostración. By definition, the n-th root of a positive real number a is the number b such that:
$$b^n = a$$

This can be rewritten as:
$$b = a^{1/n}$$

Thus, $\sqrt[n]{a}$ is simply the representation of the value $a^{1/n}$. In other words, the n-th root of a can be expressed as the power of a with a fractional exponent $1/n$.
Therefore, we have:
$$\sqrt[n]{a} = a^{1/n}$$

∎

This theorem is especially useful for working with algebraic expressions involving radicals, as it allows the application of exponent rules to manipulate and simplify them.

> **Corollary 2.2.2** For any non-negative real number a and any positive integers m, n, the following relationship holds:
> $$\sqrt[m]{\sqrt[n]{a}} = \sqrt[mn]{a}$$
> This implies that the composition of roots can be expressed as a single root of order equal to the product, simplifying nested root calculations.

Demostración. We start by expressing each root in terms of fractional exponents. The n-th root of a can be written as:
$$\sqrt[n]{a} = a^{1/n}$$

Now, taking the m-th root of this expression:
$$\sqrt[m]{\sqrt[n]{a}} = \sqrt[m]{a^{1/n}}$$

Applying the property of fractional exponents, this can be rewritten as:
$$(a^{1/n})^{1/m} = a^{(1/n) \cdot (1/m)} = a^{1/(mn)}$$

Finally, we rewrite this expression as a root of order mn:
$$a^{1/(mn)} = \sqrt[mn]{a}$$

∎

This corollary provides a straightforward way to manipulate and simplify expressions containing multiple nested radicals.

 Radicals play a significant role in various areas of mathematics, from solving equations to geometry and calculus. Understanding the properties of radicals is essential for simplifying expressions and solving complex problems involving powers and roots.

Exercise 2.4 Simplify the following radical expressions:
1. $\sqrt[3]{8x^3}$
2. $\sqrt{50} \cdot \sqrt{2}$
3. $\sqrt[4]{81y^8}$

Then, verify whether the properties of radicals hold in each case.

The study of radicals and their properties is a foundational skill for developing more advanced algebraic techniques. In later chapters, we will explore how these properties apply to solving irrational equations, analyzing functions with roots, and integrating expressions involving radicals. Moreover, radicals frequently appear in geometric formulas and the study of physical phenomena, making a solid understanding of their properties essential.

2.2.2 Simplification and Rationalization of Radicals

Definition 2.2.2 **Simplification of radicals** consists of reducing a radical to its simplest form, eliminating square factors inside the root whenever possible. **Rationalization of radicals** involves rewriting an expression that contains radicals in the denominator so that the denominator becomes a rational number. This is usually done by multiplying both the numerator and the denominator by an appropriate radical.

Simplifying and rationalizing radicals is an important step in algebra, as it allows us to work with expressions in more manageable forms, facilitating operations such as addition, subtraction, and multiplication.

■ **Example 2.5** Let's simplify the radical $\sqrt{50}$:

$$\sqrt{50} = \sqrt{25 \times 2} = \sqrt{25} \cdot \sqrt{2} = 5\sqrt{2}$$

Now, let's rationalize the fraction $\frac{3}{\sqrt{2}}$:

$$\frac{3}{\sqrt{2}} \cdot \frac{\sqrt{2}}{\sqrt{2}} = \frac{3\sqrt{2}}{2}$$

In both cases, we have simplified the radical expressions to make them more manageable. ■

Lema 2.2.2 For any non-negative real numbers $a, b \in \mathbb{R}$ and a positive integer n, the following property holds:

$$\sqrt[n]{a \cdot b} = \sqrt[n]{a} \cdot \sqrt[n]{b}$$

This property allows the simplification of products inside radicals into products of individual radicals.

This property is useful for simplifying complex expressions involving products of radicals and is foundational in algebra and analysis.

2.2 Equations with Radicals and Rationalization

Theorem 2.2.3 To rationalize a fraction of the form $\frac{a}{\sqrt{b}}$, where $a, b \in \mathbb{R}$ and $b > 0$, it is sufficient to multiply both the numerator and the denominator by $\sqrt{b}$:

$$\frac{a}{\sqrt{b}} \cdot \frac{\sqrt{b}}{\sqrt{b}} = \frac{a\sqrt{b}}{b}$$

This ensures that the denominator becomes a rational number.

Demostración. We start with the original fraction:

$$\frac{a}{\sqrt{b}}$$

The goal is to eliminate the radical from the denominator so that the denominator becomes a rational number. To do this, we multiply both the numerator and the denominator by $\sqrt{b}$:

$$\frac{a}{\sqrt{b}} \cdot \frac{\sqrt{b}}{\sqrt{b}} = \frac{a \cdot \sqrt{b}}{\sqrt{b} \cdot \sqrt{b}}$$

The denominator simplifies since $\sqrt{b} \cdot \sqrt{b} = b$:

$$\frac{a \cdot \sqrt{b}}{b} = \frac{a\sqrt{b}}{b}$$

Thus, we have eliminated the radical from the denominator and obtained an equivalent fraction with a rational denominator. ∎

This theorem provides a systematic technique for eliminating radicals from the denominator, which is particularly useful when adding or subtracting fractions containing radicals.

Corollary 2.2.4 If we have a fraction with a binomial denominator containing radicals, such as $\frac{a}{b+\sqrt{c}}$, we can rationalize it by multiplying both the numerator and the denominator by the conjugate $b - \sqrt{c}$:

$$\frac{a}{b+\sqrt{c}} \cdot \frac{b-\sqrt{c}}{b-\sqrt{c}} = \frac{a(b-\sqrt{c})}{b^2 - c}$$

This eliminates the radicals from the denominator and rewrites the fraction in a more convenient form.

Using the conjugate is a powerful technique for simplifying fractions with radicals in the denominator, enabling us to work with more accessible expressions.

> (R) Simplifying and rationalizing radicals are important tools used to solve algebraic equations, calculate limits, and simplify expressions in differential and integral calculus. These techniques help rewrite expressions into more convenient forms for analysis and resolution.

Exercise 2.5 Simplify and rationalize the following expressions:
1. $\frac{5}{\sqrt{3}}$
2. $\sqrt{72}$
3. $\frac{4}{1+\sqrt{5}}$

Then, verify if the simplified expressions allow easier operations when adding or subtracting with other radical expressions.

The process of simplifying and rationalizing radicals is essential for advancing in algebra and preparing for more complex topics, such as solving irrational equations and integrating functions involving roots. In subsequent chapters, we will explore how these techniques apply to function analysis and limit evaluation in calculus. Moreover, these tools are useful in geometric contexts, such as trigonometry, where roots frequently appear in formulas related to distances and angles.

2.2.3 Solving Radical Equations

Definition 2.2.3 A **radical equation** is an equation that contains one or more roots, typically in the form $\sqrt[n]{f(x)} = g(x)$, where $f(x)$ and $g(x)$ are algebraic functions. Solving a radical equation involves eliminating the radicals, usually by raising both sides of the equation to an appropriate power, and then solving the resulting equation.

Radical equations require careful treatment when raising both sides of the equation to a power, as this can introduce "extraneous" or false solutions. Let's look at some examples to better understand how to solve these equations.

■ **Example 2.6** Consider the equation $\sqrt{x+3} = 5$. To solve it, we first eliminate the radical by squaring both sides:

$$(\sqrt{x+3})^2 = 5^2$$

$$x+3 = 25$$

$$x = 22$$

We can verify that $x = 22$ is the solution by substituting it back into the original equation:

$$\sqrt{22+3} = \sqrt{25} = 5$$

■

Lema 2.2.3 For any equation of the form $\sqrt[n]{f(x)} = g(x)$, where $f(x)$ and $g(x)$ are functions defined on the set of real numbers and n is a positive integer, raising both sides of the equation to the power of n will result in an equivalent equation, provided the appropriate domain of $f(x)$ and $g(x)$ is considered.

This lemma provides a strategy for solving radical equations: eliminate the radical by raising both sides to the corresponding power. However, it is essential to verify the obtained solutions to avoid including those that do not satisfy the original equation.

Theorem 2.2.5 If a radical equation has the form $\sqrt[n]{f(x)} = g(x)$, and squaring or raising both sides to the n-th power results in a polynomial equation, the solution set of the original equation may contain additional roots that do not satisfy the initial equation. These are called **extraneous solutions**, and they must be individually verified.

This theorem highlights the importance of verifying each solution obtained when solving radical equations to ensure that all solutions satisfy the original equation.

Corollary 2.2.6 If an equation contains more than one radical, it may be necessary to repeat the process of raising both sides to a power multiple times until all radicals are eliminated. At each step, care must be taken to check for solutions that do not satisfy the original equation.

This corollary explains how to handle equations with multiple radicals, ensuring that all radical terms are correctly eliminated and the solutions are verified at the end.

2.3 Inequalities with Absolute Value and Their Graphical Representation

 Radical equations appear in many areas of mathematics and sciences, such as physics and engineering, where they are used to model phenomena involving quantities that grow or decrease nonlinearly. Solving these equations correctly requires a deep understanding of the properties of radicals and how to manipulate them.

> **Exercise 2.6** Solve the following radical equations and verify whether the solutions are valid:
> 1. $\sqrt{x-4} = 3$
> 2. $\sqrt[3]{2x+1} = 4$
> 3. $\sqrt{x+5} + 2 = 7$
>
> Then, explain the verification process and whether any solution is extraneous.

Studying radical equations is essential for advancing in algebra and understanding how to solve more complex problems involving roots and powers. In subsequent chapters, we will explore how these techniques apply to function analysis, integration of radical expressions, and geometric problems involving distance and optimization of areas and volumes. Moreover, these equations frequently appear in modeling phenomena in physics, chemistry, and biology, demonstrating the importance of mastering their resolution methods.

2.3 Inequalities with Absolute Value and Their Graphical Representation

2.3.1 Definition of Absolute Value Inequalities

> **Definition 2.3.1** An **absolute value inequality** is an inequality involving an expression within an absolute value. It can be written in forms such as $|f(x)| < a$, $|f(x)| \leq a$, $|f(x)| > a$, or $|f(x)| \geq a$, where $f(x)$ is a function and a is a positive real number. The absolute value represents the distance from the origin, so these inequalities define an interval or set of points on the number line.

To solve absolute value inequalities, it is necessary to split the inequality into two parts without the absolute value, allowing us to solve two equivalent linear inequalities.

■ **Example 2.7** Consider the inequality $|x-3| < 5$. To solve it, we must split it into two inequalities:

$$-5 < x - 3 < 5$$

Adding 3 to each side, we obtain:

$$-2 < x < 8$$

Thus, the solution is the interval $(-2, 8)$. This indicates that all values of x within this interval satisfy the inequality. ■

Lema 2.3.1 For any real number $a > 0$, the inequality $|x| < a$ has a solution in the interval $-a < x < a$, while the inequality $|x| > a$ has a solution in the union of intervals $x < -a$ or $x > a$.

This lemma is crucial for understanding how absolute value affects the definition of intervals in inequalities. Splitting the inequality into two simpler inequalities allows for a systematic resolution.

> **Theorem 2.3.1** If $|f(x)| \leq a$, where $a \geq 0$, then the solution lies in the interval $-a \leq f(x) \leq a$. Conversely, if $|f(x)| \geq a$, the solution lies in the intervals $f(x) \leq -a$ or $f(x) \geq a$. This implies that any absolute value inequality can be reduced to two linear inequalities.

Demostración. First, consider the inequality $|f(x)| \leq a$, with $a \geq 0$:

By the definition of absolute value, the expression $|f(x)| \leq a$ means that the distance between $f(x)$ and 0 is less than or equal to a. This translates to:

$$-a \leq f(x) \leq a$$

Thus, the solution for $|f(x)| \leq a$ is found in the interval $-a \leq f(x) \leq a$.
Next, consider the inequality $|f(x)| \geq a$:
This implies that the distance between $f(x)$ and 0 is greater than or equal to a. This translates to:

$$f(x) \leq -a \quad \text{or} \quad f(x) \geq a$$

Thus, the solution for $|f(x)| \geq a$ lies in the intervals $f(x) \leq -a$ or $f(x) \geq a$.
In both cases, it is evident that absolute value inequalities can be reduced to two linear inequalities.

This theorem provides a general method for solving any absolute value inequality by splitting the original inequality into two parts and solving each separately.

Corollary 2.3.2 If $|f(x)| < a$, where $a > 0$, then $f(x)$ is bounded by $-a$ and a. This property allows us to define an upper and lower bound for the function, which is useful for determining whether a function is restricted within certain limits.

This corollary is particularly relevant when studying bounded functions and determining intervals within which a function must remain to satisfy specific conditions.

> (R) Absolute value inequalities are useful in many contexts, especially when working with limits and continuity, as the absolute value measures distance and allows us to define neighborhoods around critical points. Understanding how to solve these inequalities is essential for studying system stability and analyzing function behavior.

Exercise 2.7 Solve the following absolute value inequalities:
1. $|2x - 1| < 3$
2. $|x + 4| \geq 7$
3. $|3x - 5| \leq 4$

Then, graphically represent the solutions obtained on the number line and explain how the absolute value affects the solution.

The study of absolute value inequalities provides fundamental tools for solving more advanced problems in algebra and calculus, such as evaluating limits and solving inequalities in multiple variables. Additionally, the ability to solve these inequalities has applications in optimization and determining function stability within a specific interval. In later chapters, we will see how these techniques are used for function analysis and determining inflection points, as well as local maxima or minima.

2.3.2 Step-by-Step Solution of Inequalities

Definition 2.3.2 An **inequality** is a mathematical expression that relates algebraic expressions using the symbols $<, \leq, >$, or $\geq$. The **solution of an inequality** involves finding the values of the variable that satisfy the inequality. This process includes simplifying the inequality, isolating

2.3 Inequalities with Absolute Value and Their Graphical Representation

the variable, and sometimes expressing the solution in terms of intervals.

To solve an inequality, it is important to follow a series of steps, including simplification, isolating the variable, and verifying the solution. Representing the solution on the number line is also essential to visualize the interval of values that satisfy the inequality.

Example 2.8 Consider the inequality $3x - 5 < 4$. To solve it, follow these steps:
1. Add 5 to both sides: $3x - 5 + 5 < 4 + 5$
2. Simplify: $3x < 9$
3. Divide both sides by 3: $x < 3$

The solution is $x < 3$, which can be represented on the number line as all values less than 3. ∎

Lema 2.3.2 If we multiply or divide both sides of an inequality by a negative number, the direction of the inequality is reversed. That is, if $a < b$ and both sides are multiplied by -1, then $-a > -b$.

This lemma is fundamental for solving inequalities because any operation involving multiplication or division by a negative number changes the direction of the inequality. This must be carefully considered when working with negative coefficients.

Theorem 2.3.3 For any inequality of the form $a < bx + c$, where $b \neq 0$, the solution is found by isolating x in a manner similar to solving equations, but with attention to the sign change rule if dividing by a negative number. The solution can be expressed as an interval.

This theorem generalizes the process of solving linear inequalities, indicating that the same steps as in equations can be followed, with attention to the signs during the process.

Corollary 2.3.4 If an inequality has the form $ax + b > 0$ and $a > 0$, then the solution is given by $x > -\frac{b}{a}$. This corollary shows how the inequality can be simplified to obtain a direct expression for the solution.

This corollary is useful for quickly identifying the solution to linear inequalities with positive coefficients, providing a direct formula for the variable's value.

(R) Inequalities are a powerful tool in solving optimization problems and modeling real-world scenarios. Understanding how to solve them step by step is fundamental for applying algebra to practical contexts, such as in economics, physics, and system design.

Exercise 2.8 Solve the following inequalities and represent the solutions on the number line:
1. $2x + 7 \geq 13$
2. $-4x + 5 < 1$
3. $\frac{x}{3} - 2 > 0$

Then, explain how the solution changes if you multiply or divide by a negative number.

Solving inequalities is essential for advancing in algebra and understanding how to represent constraints in real-world problems. In later chapters, we will explore how these techniques apply to function analysis and optimization, where inequalities allow us to define the limits within which a function reaches its maximum or minimum value. Additionally, the use of inequalities extends to differential calculus and solving problems with constraints, where understanding the set of permissible values for the solution is crucial.

2.3.3 Graphical Representation of Solutions

Definition 2.3.3 The **graphical representation of solutions** for an inequality involves representing all values that satisfy the given inequality on the number line or in the Cartesian plane.

For single-variable inequalities, a line is used to indicate the interval of solutions, and open or closed points are employed to show whether the endpoints are included.

Graphical representation of an inequality is a highly useful visual tool that allows for better interpretation and understanding of the solution set of the inequality.

■ **Example 2.9** Consider the inequality $x + 2 \leq 5$. First, solve the inequality:

$$x + 2 \leq 5$$

$$x \leq 3$$

The solution is $x \leq 3$, which represents all values less than or equal to 3. This is represented on the number line as follows:

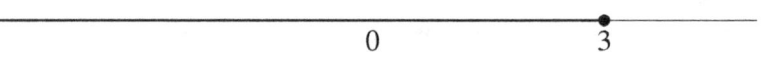

Figura 2.3.1: *Graphical representation of the solution $x \leq 3$.*

■

Lema 2.3.3 For any linear inequality of the form $ax + b < c$, where $a \neq 0$, the solution can be represented on the number line as an open or closed interval depending on the inequality sign. If the inequality is strict ($<$ or $>$), the endpoints are not included; if it is non-strict ($\leq$ or $\geq$), the endpoints are included.

This lemma provides a guideline for determining how to represent the endpoints of an interval, either with an open or closed point on the number line.

Theorem 2.3.5 The graphical representation of the solution to a two-variable inequality, such as $y < 2x + 1$, is the region of the plane below the line $y = 2x + 1$. This line can be graphed as solid or dashed depending on the type of inequality: dashed if it is strict ($<$ or $>$), and solid if it is non-strict ($\leq$ or $\geq$).

This theorem explains how to graphically represent solutions of two-variable inequalities by using a line to divide the plane into regions representing the values that satisfy the inequality.

Corollary 2.3.6 If a two-variable inequality has the form $y \geq mx + b$, then the solution includes all points on the line $y = mx + b$ and all points above it. This area can be represented as a shaded region in the Cartesian plane.

This corollary complements the theorem by describing how to identify the solution region and highlight the area that satisfies the inequality.

(R) Graphically representing solutions of an inequality not only facilitates understanding the solution set but also helps analyze intersections of regions, especially when solving systems of inequalities. These representations are fundamental in optimization problems and linear programming.

Exercise 2.9 Graphically represent the solutions to the following inequalities:
1. $x - 4 > -2$
2. $2x + y \leq 6$ (in the Cartesian plane)

3. $|x-3| \leq 4$

Then, verify if the graphical representation matches the algebraic solution obtained.

The graphical representation of solutions allows for a better visualization of the values that satisfy inequalities, and its analysis is fundamental for solving optimization problems and studying constraints in various contexts. In subsequent chapters, we will explore how these techniques apply to solving systems of inequalities and analyzing functions, where graphical representation helps determine regions of interest, critical points, and areas of optimization. Additionally, graphical representation is essential for understanding calculus concepts, such as lateral limits and continuity in function analysis.

2.4 Solved Exercises

Exercise 2.10 Solve the polynomial equation $x^4 - 5x^2 + 4 = 0$ using factorization and find all its roots.

Demostración. First, make a substitution to simplify the equation. Let $y = x^2$, which transforms the equation into:

$$y^2 - 5y + 4 = 0$$

Now, factorize the quadratic equation:

$$y^2 - 5y + 4 = (y-1)(y-4) = 0$$

Solve for y:

$$y = 1 \quad \text{or} \quad y = 4$$

Since $y = x^2$, revert to the variable x:

$$x^2 = 1 \quad \Rightarrow \quad x = \pm 1$$

$$x^2 = 4 \quad \Rightarrow \quad x = \pm 2$$

Therefore, the roots of the original equation are:

$$x = -2, -1, 1, 2$$

∎

Exercise 2.11 Simplify the expression $\frac{\sqrt{x}+3}{\sqrt{x}-3}$ by rationalizing the denominator.

Capítulo 2. Equations and Inequalities

Demostración. To rationalize the denominator of $\frac{\sqrt{x}+3}{\sqrt{x}-3}$, multiply both the numerator and denominator by the conjugate of the denominator, $\sqrt{x}+3$:

$$\frac{\sqrt{x}+3}{\sqrt{x}-3} \cdot \frac{\sqrt{x}+3}{\sqrt{x}+3} = \frac{(\sqrt{x}+3)^2}{(\sqrt{x}-3)(\sqrt{x}+3)}$$

Calculate the numerator and denominator separately:

$$(\sqrt{x}+3)^2 = x + 6\sqrt{x} + 9$$

$$(\sqrt{x}-3)(\sqrt{x}+3) = x - 9$$

Thus, the simplified expression is:

$$\frac{x + 6\sqrt{x} + 9}{x - 9}$$

∎

Exercise 2.12 Solve the radical equation $\sqrt{3x-4} = x - 2$ and verify if the solutions are real.

Demostración. First, square both sides of the equation to eliminate the radical:

$$(\sqrt{3x-4})^2 = (x-2)^2$$

$$3x - 4 = x^2 - 4x + 4$$

Rearrange all terms to obtain a quadratic equation:

$$x^2 - 7x + 8 = 0$$

Use the quadratic formula to solve:

$$x = \frac{-b \pm \sqrt{b^2 - 4ac}}{2a}$$

where $a = 1$, $b = -7$, and $c = 8$. Substitute the values:

$$x = \frac{-(-7) \pm \sqrt{(-7)^2 - 4 \cdot 1 \cdot 8}}{2 \cdot 1}$$

$$x = \frac{7 \pm \sqrt{49 - 32}}{2}$$

2.4 Solved Exercises

$$x = \frac{7 \pm \sqrt{17}}{2}$$

This gives two possible solutions:

$$x_1 = \frac{7 + \sqrt{17}}{2}, \quad x_2 = \frac{7 - \sqrt{17}}{2}$$

Verify if both solutions satisfy the original equation. For $x_2 = \frac{7-\sqrt{17}}{2}$, since the value is less than 2, $\sqrt{3x-4}$ cannot be negative, so this solution is invalid.
The only real solution is:

$$x = \frac{7 + \sqrt{17}}{2}$$

∎

Exercise 2.13 Determine all values of x that satisfy the inequality $|2x+5| \geq 7$. Represent the solution on the number line.

Demostración. To solve $|2x+5| \geq 7$, consider two cases due to the definition of absolute value:

$$2x+5 \geq 7 \quad \text{or} \quad 2x+5 \leq -7$$

Solve each inequality:
- $2x+5 \geq 7$

$$2x \geq 2$$

$$x \geq 1$$

- $2x+5 \leq -7$

$$2x \leq -12$$

$$x \leq -6$$

Thus, the solution is:

$$x \leq -6 \quad \text{or} \quad x \geq 1$$

On the number line, the solution is represented as two intervals: $(-\infty, -6] \cup [1, \infty)$. ∎

Exercise 2.14 Use synthetic division to check if $x = -2$ is a root of $x^3 + 4x^2 - x - 6 = 0$.

Demostración. To check if $x = -2$ is a root of $x^3 + 4x^2 - x - 6 = 0$ using synthetic division, perform the following:

Coefficients: $1, 4, -1, -6$

$$\begin{array}{r|rrrr} -2 & 1 & 4 & -1 & -6 \\ & & -2 & -4 & 10 \\ \hline & 1 & 2 & -5 & 4 \end{array}$$

The remainder is 4, indicating that $x = -2$ is not a root of the equation since the remainder is not zero.

Thus, $x = -2$ is not a root of the polynomial $x^3 + 4x^2 - x - 6$. ∎

2.5 Proposed Exercises

2.5.1 Solving Polynomial Equations

Exercise 2.15 Solve the equation $x^3 - 6x^2 + 11x - 6 = 0$ using factorization.

Exercise 2.16 Find the roots of the equation $x^4 - 16 = 0$ and classify them as real or complex.

Exercise 2.17 Prove that the equation $x^5 - 2x^3 + x = 0$ has at least one real root.

Exercise 2.18 Use the synthetic division method to verify if $x = 2$ is a root of $x^3 - 4x^2 + x + 6 = 0$.

Exercise 2.19 Find the irreducible quadratic factor of $x^4 + 6x^2 + 9$.

2.5.2 Equations with Radicals and Rationalization

Exercise 2.20 Solve the equation $\sqrt{x+3} + 2 = x$ and verify if the solutions are real.

Exercise 2.21 Simplify the expression $\frac{\sqrt{x+2}}{\sqrt{x-2}}$ by multiplying the numerator and denominator by the conjugate.

Exercise 2.22 Determine all the values of x that satisfy the equation $\sqrt{2x+5} = x - 1$.

Exercise 2.23 Solve the equation $\sqrt[3]{x+4} = 2$ and represent the solution on the number line.

Exercise 2.24 Find the solution to the equation $\sqrt{3x-4} = x - 2$ and verify if it is a valid solution.

2.5.3 Inequalities with Absolute Value and Their Graphical Representation

2.5 Proposed Exercises

Exercise 2.25 Solve the inequality $|3x-1| > 5$ and represent the solution on the number line.

Exercise 2.26 Determine all the values of x that satisfy $|2x+3| \leq 7$.

Exercise 2.27 Solve the inequality $|x^2 - 4| < 3$ and describe the solution set.

Exercise 2.28 Graphically represent the solution of the inequality $|x+2| > 4$ on the number line.

Exercise 2.29 Explain why the inequality $|x-1| \geq -2$ has all real numbers as its solution.

3. Complex Number System

3.1 Representation of Complex Numbers in the Plane

3.1.1 Rectangular Form of a Complex Number

Definition 3.1.1 The **rectangular form** of a complex number is expressed as $z = a + bi$, where a and b are real numbers, and i is the imaginary unit, defined as $i^2 = -1$. The term a is called the **real part**, and b is called the **imaginary part** of the complex number.

The rectangular form is the most common way to represent complex numbers. It provides a convenient method for performing operations like the addition and subtraction of complex numbers.

■ **Example 3.1** Consider the complex number $z = 3 + 4i$. In this case, the real part is $a = 3$, and the imaginary part is $b = 4$. The complex number can be graphically represented in the complex plane (or Argand plane): ■

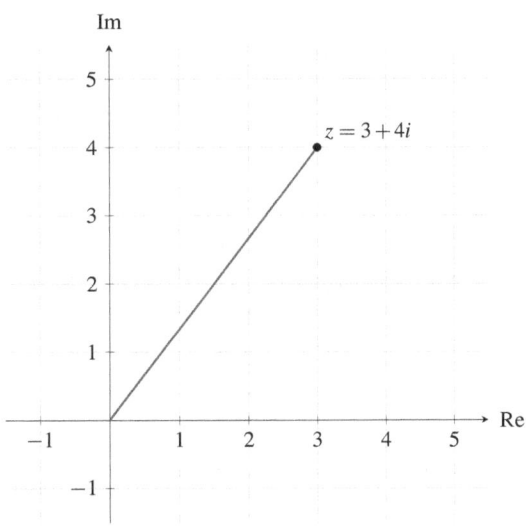

Figura 3.1.1: *Graphical representation of the complex number $z = 3 + 4i$ in the complex plane.*

Lema 3.1.1 The sum of two complex numbers $z_1 = a+bi$ and $z_2 = c+di$ is given by the sum of their real and imaginary parts:

$$z_1 + z_2 = (a+c) + (b+d)i$$

This lemma shows that the addition of complex numbers is performed by independently adding the real parts and the imaginary parts. This makes addition in the set of complex numbers both commutative and associative.

Theorem 3.1.1 For two complex numbers $z_1 = a+bi$ and $z_2 = c+di$, the product is given by:

$$z_1 \cdot z_2 = (ac - bd) + (ad + bc)i$$

Demostración. Consider two complex numbers $z_1 = a+bi$ and $z_2 = c+di$, where $a,b,c,d \in \mathbb{R}$. The product of z_1 and z_2 is defined as:

$$z_1 \cdot z_2 = (a+bi)(c+di)$$

Using the distributive property, we expand each term:

$$(a+bi)(c+di) = ac + adi + bci + bdi^2$$

Since $i^2 = -1$, we simplify:

$$ac + adi + bci - bd$$

Grouping the real and imaginary terms, we get:

$$z_1 \cdot z_2 = (ac - bd) + (ad + bc)i$$

Thus, the product of two complex numbers is:

$$z_1 \cdot z_2 = (ac - bd) + (ad + bc)i$$

∎

The above theorem describes how to multiply two complex numbers using the distributive property and the definition of the imaginary unit i, where $i^2 = -1$. The result is also expressed in rectangular form.

Corollary 3.1.2 The **conjugate** of a complex number $z = a+bi$ is denoted as $\bar{z} = a-bi$. The product of a complex number with its conjugate always results in a non-negative real number:

$$z \cdot \bar{z} = a^2 + b^2$$

Demostración. Let $z = a+bi$, where $a,b \in \mathbb{R}$. Its conjugate is given by $\bar{z} = a-bi$. Multiply z by its conjugate:

$$z \cdot \bar{z} = (a+bi)(a-bi)$$

3.1 Representation of Complex Numbers in the Plane

Using the distributive property:

$$(a+bi)(a-bi) = a^2 - abi + abi - b^2i^2$$

The terms $-abi$ and $+abi$ cancel each other:

$$a^2 - b^2i^2$$

Since $i^2 = -1$, we have:

$$a^2 - b^2(-1) = a^2 + b^2$$

Thus, the product of a complex number with its conjugate is:

$$z \cdot \bar{z} = a^2 + b^2$$

The result $a^2 + b^2$ is always a non-negative real number, as $a^2 \geq 0$ and $b^2 \geq 0$. ■

This corollary is particularly useful in the division of complex numbers, as the conjugate allows rationalization.°f the denominator, simplifying the resulting expression.

> (R) The rectangular representation of complex numbers facilitates basic arithmetic operations and provides a way to geometrically interpret complex numbers in the Argand plane. Additionally, the rectangular form bridges algebraic concepts with their graphical representation, which is essential in physical applications and signal analysis.

> Exercise 3.1 Perform the following operations with complex numbers and represent the results graphically in the complex plane:
> 1. Addition: $(2+3i)+(1-4i)$
> 2. Multiplication: $(1+2i) \cdot (3-i)$
> 3. Conjugate and multiplication: Find the conjugate of $z = 4 + 5i$ and calculate $z \cdot \bar{z}$
>
> Then, graphically represent the results and verify their algebraic properties.

The study of the rectangular form of complex numbers is fundamental for understanding basic operations in the set of complex numbers and their representation in the plane. Later, we will explore how to convert from rectangular form to polar form, and how these representations are applied in the **De Moivre's formula** to calculate powers and roots of complex numbers. The connection between algebraic and geometric forms of complex numbers allows a deeper understanding of their properties and applications in fields such as electrical engineering, quantum physics, and signal analysis.

3.1.2 Representation in the Argand Plane

> **Definition 3.1.2** The **Argand plane**, also known as the **complex plane**, is a two-dimensional graphical representation where complex numbers are expressed as points or vectors. In this plane, the horizontal axis (**real axis**) represents the real part of the complex number, while the vertical axis (**imaginary axis**) represents the imaginary part. Thus, a complex number $z = a + bi$ is represented as the point (a, b) in the plane.

The graphical representation of complex numbers in the Argand plane facilitates understanding their properties and operations such as addition, subtraction, and multiplication. It also allows visualization of concepts like magnitude and argument of a complex number.

■ **Example 3.2** Consider the complex number $z = -2 + 3i$. To represent its modulus and argument, we calculate:

$$|z| = \sqrt{(-2)^2 + 3^2} = \sqrt{4+9} = \sqrt{13} \approx 3{,}61$$

For the argument, note that the complex number lies in the second quadrant, so:

$$\theta = \pi - \tan^{-1}\left(\frac{3}{2}\right) \approx 123{,}69°$$

In radians:

$$\theta \approx \pi - \tan^{-1}\left(\frac{3}{2}\right) \approx 2{,}16 \text{ radians.}$$

In the following graph, the complex number is represented as a vector from the origin to the point $(-2, 3)$.

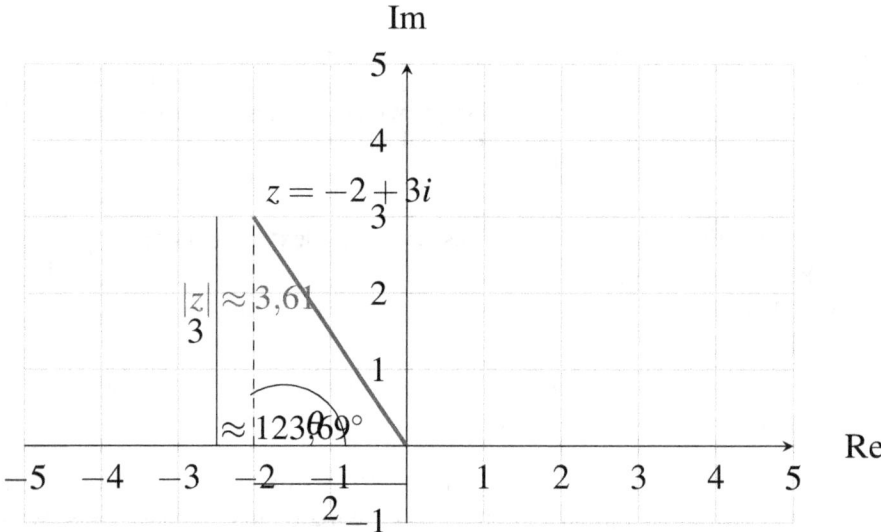

Figura 3.1.2: *Representation of the complex number $z = -2 + 3i$ in the Argand plane, showing its modulus and argument.*

■

Lema 3.1.2 The magnitude (or modulus) of a complex number $z = a + bi$ is determined as the distance from the origin to the point (a, b) in the Argand plane, given by the formula:

$$|z| = \sqrt{a^2 + b^2}$$

The magnitude of a complex number is analogous to the length of a vector in two-dimensional space and is essential for understanding properties such as the distance between complex numbers or the norm of a complex number.

3.1 Representation of Complex Numbers in the Plane

> **Theorem 3.1.3** The **argument** of a complex number $z = a + bi$ is the angle θ formed by the vector corresponding to the complex number with the positive real axis. It is calculated using the inverse tangent function:
> $$\theta = \tan^{-1}\left(\frac{b}{a}\right), \quad \text{where } a \neq 0.$$

Demostración. Consider the complex number $z = a + bi$, where $a, b \in \mathbb{R}$, represented in the complex plane. In this representation, the complex number z appears as the point or vector (a, b) in the plane. The argument θ is the angle formed by the vector (a, b) with the positive real axis. Using trigonometry, we have the relationship to compute the angle in terms of the opposite side and the adjacent side of the right triangle formed:

$$\tan(\theta) = \frac{b}{a}, \quad \text{where } a \neq 0.$$

To isolate θ, apply the inverse tangent function:

$$\theta = \tan^{-1}\left(\frac{b}{a}\right).$$

This formula allows us to determine the angle θ based on the real and imaginary components a and b of the complex number. ∎

The argument is an important feature in the polar form of complex numbers, as it defines their orientation in the plane. It is useful in operations like multiplication and division, which involve angles.

> **Corollary 3.1.4** If z is a complex number, then the complex conjugate $\bar{z} = a - bi$ has the same magnitude as z, but its argument is the opposite:
>
> If θ is the argument of z, then the argument of $\bar{z}$ is $-\theta$.

Demostración. Consider the complex number $z = a + bi$, with $a, b \in \mathbb{R}$. In the Argand plane, z is represented as the point (a, b).
The argument θ of z is the angle formed by the vector (a, b) with the positive real axis:

$$\theta = \tan^{-1}\left(\frac{b}{a}\right), \quad \text{where } a \neq 0.$$

Now, consider the conjugate $\bar{z} = a - bi$. In the Argand plane, $\bar{z}$ is represented as the point $(a, -b)$. The argument of $\bar{z}$, denoted as $-\theta$, is the angle formed by the vector $(a, -b)$ with the positive real axis:

$$-\theta = \tan^{-1}\left(\frac{-b}{a}\right).$$

Thus, the argument of the conjugate is the opposite of the original complex number's argument, $-\theta$.

Moreover, the magnitudes of z and $\bar{z}$ are identical since:

$$|z| = \sqrt{a^2 + b^2} = |\bar{z}|.$$

This demonstrates that the conjugate has the same magnitude but the opposite argument. ∎

The conjugate reflects the complex number across the real axis, which is important in rationalizing denominators and determining the square roots of complex numbers.

> R The representation of complex numbers in the Argand plane helps visualize algebraic operations and better understand properties such as modulus and argument. This graphical representation has applications in physics and engineering, particularly in signal theory and electrical circuit analysis.

> **Exercise 3.2** Represent the following complex numbers in the Argand plane, and determine their magnitude and argument:
> 1. $z_1 = 1 + i$
> 2. $z_2 = -3 + 4i$
> 3. $z_3 = -2 - 5i$
>
> Then, calculate the conjugate of each number and graph both the original complex number and its conjugate.

Representation in the Argand plane not only aids in visualizing complex numbers but also provides a strong foundation for advanced operations, such as multiplication in polar form and the application of **De Moivre's theorem** to calculate powers and roots of complex numbers. In subsequent chapters, we will explore how polar form and the geometric properties of complex numbers simplify more intricate calculations and help explain periodic phenomena in physics and other scientific disciplines.

3.1.3 Modulus and Argument of a Complex Number

> **Definition 3.1.3** The **modulus** of a complex number $z = a + bi$, denoted as $|z|$, is the distance of the complex number from the origin in the Argand plane and is calculated using the formula:
>
> $$|z| = \sqrt{a^2 + b^2}$$
>
> The **argument** of a complex number $z = a + bi$, denoted as $\arg(z)$, is the angle θ that the vector representing the complex number makes with the positive real axis. This angle is determined using:
>
> $$\theta = \tan^{-1}\left(\frac{b}{a}\right)$$
>
> when $a \neq 0$. To correctly define the argument, it is important to consider the quadrant in which the complex number is located.

The modulus and argument allow a complex number to be represented in its **polar form**, which is particularly useful for performing multiplications, divisions, and working with powers and roots.

■ **Example 3.3** Consider the complex number $z = 3 + 4i$. To calculate its modulus and argument:

$$|z| = \sqrt{3^2 + 4^2} = \sqrt{9 + 16} = \sqrt{25} = 5$$

3.1 Representation of Complex Numbers in the Plane

For the argument:

$$\theta = \tan^{-1}\left(\frac{4}{3}\right) \approx 0{,}93 \text{ radians.}$$

By plotting the complex number on the Argand plane, we can visualize the modulus as the vector's length and the argument as the angle it makes with the real axis.

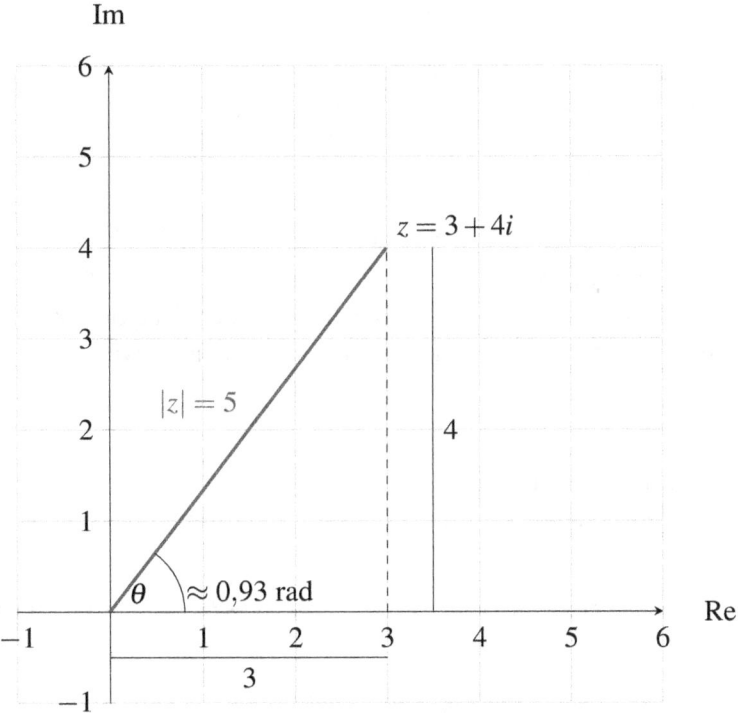

Figura 3.1.3: *Representation of the complex number $z = 3 + 4i$ with modulus $|z| = 5$ and argument $\theta \approx 0{,}93$ radians.*

■

Lema 3.1.3 For any complex number $z = a + bi$, the modulus satisfies the triangular property:

$$|z_1 + z_2| \leq |z_1| + |z_2|$$

where $z_1, z_2 \in \mathbb{C}$. This property is analogous to the triangular inequality in vector spaces and ensures that the length of the sum of two vectors does not exceed the sum of their lengths.

This lemma is fundamental in complex analysis and geometry in the complex plane, as it establishes a key relationship between the sum of complex numbers and their moduli.

Theorem 3.1.5 If a complex number z is given in polar form as $z = r(\cos\theta + i\sin\theta)$, then the modulus of z is r, and the argument is θ. This form is known as the **polar form** of a complex number and simplifies many operations, such as multiplication and division.

Demostración. Consider the complex number z represented in polar form as:

$$z = r(\cos\theta + i\sin\theta),$$

where $r \geq 0$ is the modulus and θ is the argument of the complex number z.

The modulus r of z is defined as the distance from the origin to the point in the Argand plane, which corresponds to:

$$|z| = \sqrt{a^2 + b^2},$$

where a and b are the real and imaginary parts of the complex number, respectively. In the polar form, we have:

$$a = r\cos\theta, \quad b = r\sin\theta.$$

Substituting into the modulus expression:

$$|z| = \sqrt{(r\cos\theta)^2 + (r\sin\theta)^2} = \sqrt{r^2(\cos^2\theta + \sin^2\theta)}.$$

Since $\cos^2\theta + \sin^2\theta = 1$, we obtain:

$$|z| = \sqrt{r^2} = r.$$

This proves that the modulus of z is r.

Moreover, the argument of z, denoted by θ, is the angle that the vector corresponding to z forms with the positive real axis. In the polar form, this angle is explicitly given as θ. Thus, the argument of z is θ.

■

The polar form is especially useful when calculating powers or roots of complex numbers, as multiplying complex numbers in polar form involves simply adding their arguments and multiplying their moduli.

Corollary 3.1.6 If $z_1 = r_1(\cos\theta_1 + i\sin\theta_1)$ and $z_2 = r_2(\cos\theta_2 + i\sin\theta_2)$, then the product of z_1 and z_2 is given by:

$$z_1 z_2 = r_1 r_2 \left(\cos(\theta_1 + \theta_2) + i\sin(\theta_1 + \theta_2)\right)$$

This corollary shows that when multiplying two complex numbers, the moduli are multiplied, and the arguments are added.

Demostración. To prove this corollary, consider the complex numbers z_1 and z_2 in polar form:

$$z_1 = r_1(\cos\theta_1 + i\sin\theta_1), \quad z_2 = r_2(\cos\theta_2 + i\sin\theta_2).$$

The product of z_1 and z_2 is obtained by multiplying the two expressions:

$$z_1 z_2 = [r_1(\cos\theta_1 + i\sin\theta_1)][r_2(\cos\theta_2 + i\sin\theta_2)].$$

Using the distributive property:

$$z_1 z_2 = r_1 r_2 \left[(\cos\theta_1 \cos\theta_2 - \sin\theta_1 \sin\theta_2) + i(\cos\theta_1 \sin\theta_2 + \sin\theta_1 \cos\theta_2)\right].$$

By applying the trigonometric identities for the sum of angles:

$$\cos(\theta_1 + \theta_2) = \cos\theta_1 \cos\theta_2 - \sin\theta_1 \sin\theta_2,$$

$$\sin(\theta_1 + \theta_2) = \cos\theta_1 \sin\theta_2 + \sin\theta_1 \cos\theta_2,$$

we can write:

$$z_1 z_2 = r_1 r_2 \left(\cos(\theta_1 + \theta_2) + i\sin(\theta_1 + \theta_2)\right).$$

This demonstrates that when multiplying two complex numbers in polar form, the moduli are multiplied, and the arguments are added.

■

This corollary forms the basis of **De Moivre's theorem**, which will be used in subsequent chapters to efficiently calculate powers and roots of complex numbers.

> (R) Representing complex numbers through modulus and argument allows for more intuitive operations, especially when dealing with trigonometric and geometric properties. This representation is also fundamental in signal theory and electrical engineering applications.

> **Exercise 3.3** For each of the following complex numbers, determine the modulus, the argument, and represent the complex number graphically on the Argand plane:
> 1. $z_1 = -1 + i\sqrt{3}$
> 2. $z_2 = 4 - 4i$
> 3. $z_3 = -3 - 2i$
>
> Then, write each complex number in its polar form.

Understanding the modulus and argument of a complex number is key to representing it in polar form and performing operations such as multiplication and division more efficiently. In upcoming chapters, we will explore additional applications, such as **De Moivre's theorem**, which enables us to calculate powers and roots of complex numbers, with significant applications in solving periodic problems and analyzing oscillatory signals.

3.2 Operations with Complex Numbers

3.2.1 Addition and Subtraction of Complex Numbers

> **Definition 3.2.1** Let $z_1 = a + bi$ and $z_2 = c + di$ be two complex numbers. The **sum** of z_1 and z_2 is defined as:
>
> $$z_1 + z_2 = (a+c) + (b+d)i.$$
>
> The **difference** of z_1 and z_2 is defined as:
>
> $$z_1 - z_2 = (a-c) + (b-d)i.$$

The addition and subtraction of complex numbers are simple operations performed component-wise, as shown in the following example.

■ **Example 3.4** Given $z_1 = 3 + 4i$ and $z_2 = 1 - 2i$, let us compute their sum and difference:

$$z_1 + z_2 = (3+1) + (4-2)i = 4 + 2i,$$

$$z_1 - z_2 = (3-1) + (4+2)i = 2 + 6i.$$

■

> (R) The addition and subtraction of complex numbers are closed operations, meaning the result of adding or subtracting two complex numbers is always another complex number.

Lema 3.2.1 Let $z_1 = a + bi$ and $z_2 = c + di$ be two complex numbers. The addition of complex numbers is commutative and associative, i.e.:
- $z_1 + z_2 = z_2 + z_1$.
- $(z_1 + z_2) + z_3 = z_1 + (z_2 + z_3)$ for any $z_3 \in \mathbb{C}$.

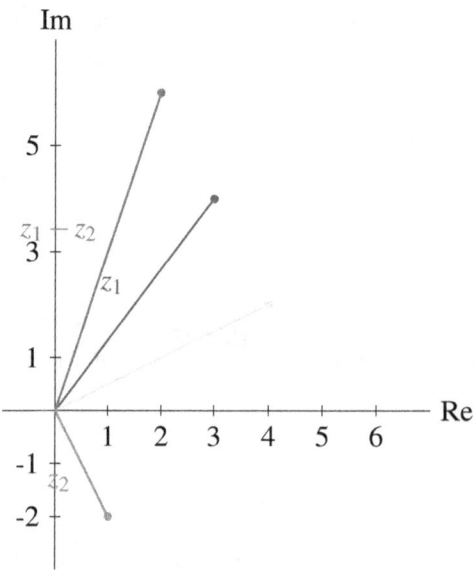

Figura 3.2.1: *Graphical representation of z_1, z_2, $z_1 + z_2$, and $z_1 - z_2$ in the complex plane.*

The proof of this lemma can be carried out by verifying each case using the definition of complex number addition.

Demostración. Let $z_1 = a + bi$, $z_2 = c + di$, and $z_3 = e + fi$ be three complex numbers.

- For commutativity, we have:

$$z_1 + z_2 = (a+c) + (b+d)i = (c+a) + (d+b)i = z_2 + z_1.$$

- For associativity, we have:

$$(z_1 + z_2) + z_3 = [(a+c) + (b+d)i] + (e+fi) = (a+c+e) + (b+d+f)i,$$

$$z_1 + (z_2 + z_3) = (a+bi) + [(c+e) + (d+f)i] = (a+c+e) + (b+d+f)i.$$

In both cases, the result is the same. ∎

The importance of commutativity and associativity lies in allowing us to reorganize and combine terms in sums of complex numbers without altering the result, which is useful for simplifying expressions.

Theorem 3.2.1 For any complex number $z = a + bi$, it holds that:

$$z + 0 = z,$$

where 0 is the additive identity element in $\mathbb{C}$.

Demostración. Let $z = a + bi$, and consider the complex number $0 = 0 + 0i$. Adding the two, we get:

$$z + 0 = (a+0) + (b+0)i = a + bi = z.$$

Thus, 0 is the additive identity element in the set of complex numbers. ∎

3.2 Operations with Complex Numbers

Corollary 3.2.2 For any complex number $z = a + bi$, there exists a complex number $-z = -a - bi$ such that:

$$z + (-z) = 0.$$

This corollary shows the existence of the additive inverse of each complex number, which is key to defining subtraction in the set of complex numbers.

■ **Example 3.5** Let $z = 2 - 3i$. Its additive inverse is $-z = -2 + 3i$, and we can verify that:

$$z + (-z) = (2 - 3i) + (-2 + 3i) = 0.$$

■

Exercise 3.4 Let $z_1 = 4 + 7i$ and $z_2 = -2 - 5i$. Compute $z_1 + z_2$, $z_1 - z_2$, and find the additive inverse of z_2.

These exercises allow the reader to practice adding and subtracting complex numbers and reinforce understanding of the properties discussed so far. Next, we will study the multiplication of complex numbers and its properties, leading to a deeper exploration of the **De Moivre's Theorem**.

3.2.2 Multiplication and Division in Rectangular Form

The multiplication and division of complex numbers in rectangular form are essential operations for working with this type of numbers. Below, each operation is defined, and some of its key properties are explored.

Definition 3.2.2 Let $z_1 = a + bi$ and $z_2 = c + di$ be two complex numbers. The **multiplication** of z_1 and z_2 is defined as:

$$z_1 \cdot z_2 = (a + bi)(c + di) = (ac - bd) + (ad + bc)i.$$

The multiplication of complex numbers in rectangular form is based on the distributive property, similar to the multiplication of binomials in real numbers, but considering the property $i^2 = -1$.

■ **Example 3.6** Let us multiply the complex numbers $z_1 = 2 + 3i$ and $z_2 = 1 - 4i$:

$$z_1 \cdot z_2 = (2 + 3i)(1 - 4i) = (2 \cdot 1 - 3 \cdot 4) + (2 \cdot -4 + 3 \cdot 1)i = -10 - 5i.$$

■

(R) The multiplication of complex numbers is commutative and associative, meaning $z_1 \cdot z_2 = z_2 \cdot z_1$ and $(z_1 \cdot z_2) \cdot z_3 = z_1 \cdot (z_2 \cdot z_3)$ for any $z_1, z_2, z_3 \in \mathbb{C}$. This allows us to rearrange the factors without altering the result.

Lema 3.2.2 For any complex number $z = a + bi$, its **magnitude** is denoted by $|z|$ and is defined as:

$$|z| = \sqrt{a^2 + b^2}.$$

The magnitude of a complex number is fundamental when studying the multiplication and division of complex numbers, as it helps us understand how these operations affect the size of the resulting complex number.

> **Theorem 3.2.3** Let $z_1 = a+bi$ and $z_2 = c+di$ be two complex numbers. The magnitude of their product is the product of their magnitudes, i.e.:
>
> $$|z_1 \cdot z_2| = |z_1| \cdot |z_2|.$$

Demostración. Let $z_1 = a+bi$ and $z_2 = c+di$. Then:

$$z_1 \cdot z_2 = (a+bi)(c+di) = (ac-bd)+(ad+bc)i.$$

Calculating the magnitude of the product:

$$|z_1 \cdot z_2| = \sqrt{(ac-bd)^2 + (ad+bc)^2}.$$

On the other hand:

$$|z_1| = \sqrt{a^2+b^2}, \quad |z_2| = \sqrt{c^2+d^2}.$$

Multiplying the magnitudes, we have:

$$|z_1| \cdot |z_2| = \sqrt{a^2+b^2} \cdot \sqrt{c^2+d^2} = \sqrt{(a^2+b^2)(c^2+d^2)}.$$

Expanding the product results in equality:

$$|z_1 \cdot z_2| = |z_1| \cdot |z_2|.$$

∎

This result is very useful for calculating the magnitude of a complex number resulting from a product, simplifying the process.

> **Corollary 3.2.4** If $|z_1| = 1$ and $|z_2| = 1$, then $|z_1 \cdot z_2| = 1$. In particular, the multiplication of complex numbers with magnitude 1 produces another complex number with magnitude 1.

> **Definition 3.2.3** The **division** of two complex numbers $z_1 = a+bi$ and $z_2 = c+di$ (with $z_2 \neq 0$) is defined as:
>
> $$\frac{z_1}{z_2} = \frac{(a+bi)(c-di)}{c^2+d^2} = \frac{ac+bd}{c^2+d^2} + \frac{bc-ad}{c^2+d^2}i.$$

The division of complex numbers in rectangular form is based on multiplying the numerator and denominator by the conjugate of the denominator, which eliminates the imaginary term in the denominator.

■ **Example 3.7** Let us divide the complex numbers $z_1 = 4+2i$ and $z_2 = 1-i$:

$$\frac{4+2i}{1-i} = \frac{(4+2i)(1+i)}{(1-i)(1+i)} = \frac{(4+2i+4i-2)}{1+1} = \frac{2+6i}{2} = 1+3i.$$

■

> **Exercise 3.5** Calculate the quotient $\frac{3-i}{2+i}$ and verify that the result obtained has the same magnitude as the quotient of the magnitudes of the complex numbers. ■

The process of multiplying and dividing complex numbers is directly connected to the concept of the conjugate, which simplifies the handling of imaginary terms in these operations.

3.2 Operations with Complex Numbers

 The use of the conjugate in the division of complex numbers facilitates the elimination of the imaginary part of the denominator, allowing the quotient to be expressed as a complex number in rectangular form.

This approach to multiplication and division in rectangular form provides a solid foundation for exploring more advanced representations of complex numbers, such as the polar form and the application of **De Moivre's Theorem**, which will be covered in the next section.

3.2.3 Conjugate of a Complex Number

Definition 3.2.4 The **conjugate** of a complex number $z = a + bi$ is denoted as $\bar{z}$ and is defined as the complex number $\bar{z} = a - bi$, where a and b are real numbers. The conjugate of z is obtained by changing the sign of the imaginary part.

The conjugate of a complex number is an important tool in many algebraic operations, such as rationalizing denominators, and it has an interesting geometric interpretation as a reflection through the real axis in the complex plane.

■ **Example 3.8** Consider the complex number $z = 3 - 4i$. The conjugate of this number is:

$$\bar{z} = 3 + 4i$$

Plotting the complex number z and its conjugate $\bar{z}$ in the Argand plane, we can see that they are symmetric with respect to the real axis.

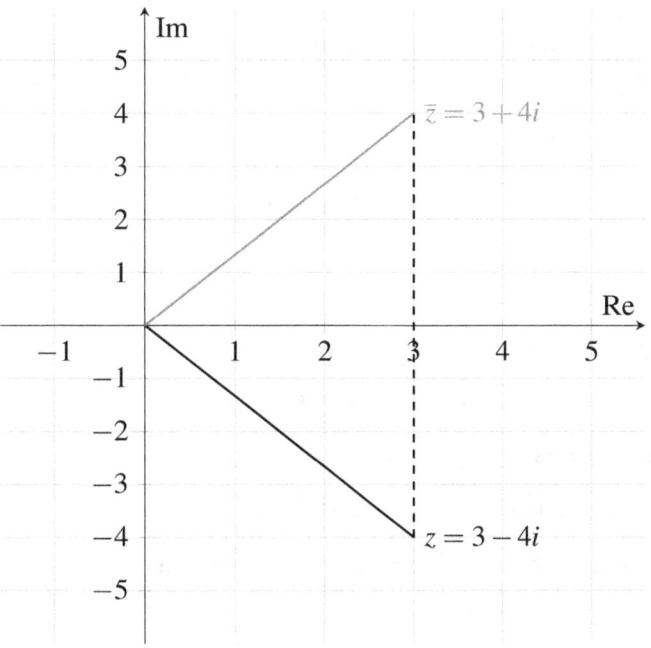

Figura 3.2.2: *Graphical representation of the complex number $z = 3 - 4i$ and its conjugate $\bar{z} = 3 + 4i$ in the Argand plane.*

■

Lema 3.2.3 For any complex number $z = a + bi$, the product of z and its conjugate $\bar{z}$ is always a non-negative real number:

$$z \cdot \bar{z} = (a + bi)(a - bi) = a^2 + b^2$$

This value corresponds to the square of the magnitude of z.

Demostración. Consider the complex number $z = a + bi$, where $a, b \in \mathbb{R}$. Its conjugate is $\bar{z} = a - bi$. The product of z and its conjugate is:

$$z \cdot \bar{z} = (a+bi)(a-bi).$$

Applying the distributive property (also known as "binomial expansion"):

$$(a+bi)(a-bi) = a^2 - (bi)^2.$$

Since $(bi)^2 = b^2 i^2$ and $i^2 = -1$, we get:

$$(bi)^2 = -b^2.$$

Thus:

$$(a+bi)(a-bi) = a^2 - (-b^2) = a^2 + b^2.$$

We observe that $a^2 + b^2 \geq 0$ for any $a, b \in \mathbb{R}$, which is always a non-negative real number. Moreover, this value is precisely the square of the magnitude of z, as:

$$|z| = \sqrt{a^2 + b^2}, \quad \text{and hence} \quad |z|^2 = a^2 + b^2.$$

∎

This lemma is fundamental because it shows how the product of a complex number and its conjugate eliminates the imaginary part, resulting in a real number. This property is used, for example, to rationalize fractions with complex numbers.

> **Theorem 3.2.5** If z_1 and z_2 are complex numbers, then the conjugate of their product equals the product of their conjugates:
>
> $$\overline{z_1 z_2} = \overline{z_1} \cdot \overline{z_2}.$$
>
> This property extends to any number of factors and shows that the conjugate distributes over multiplication.

Demostración. Let $z_1 = a + bi$ and $z_2 = c + di$, where $a, b, c, d \in \mathbb{R}$.
First, compute the product $z_1 z_2$:

$$z_1 z_2 = (a+bi)(c+di) = (ac - bd) + (ad + bc)i.$$

Now, find the conjugate of $z_1 z_2$:

$$\overline{z_1 z_2} = \overline{(ac - bd) + (ad + bc)i} = (ac - bd) - (ad + bc)i.$$

On the other hand, compute the conjugate of each factor and then their product:

$$\overline{z_1} = a - bi, \quad \overline{z_2} = c - di.$$

The product of the conjugates is:

$$\overline{z_1} \cdot \overline{z_2} = (a - bi)(c - di) = (ac - bd) - (ad + bc)i.$$

We observe that:

$$\overline{z_1 z_2} = \overline{z_1} \cdot \overline{z_2},$$

which proves the property.

∎

3.2 Operations with Complex Numbers

This property is important when working with complex expressions and simplifies handling products of complex numbers in various applications, such as signal processing and quantum physics.

Corollary 3.2.6 The conjugate of the sum of two complex numbers z_1 and z_2 is equal to the sum of their conjugates:

$$\overline{z_1 + z_2} = \overline{z_1} + \overline{z_2}.$$

Demostración. Let $z_1 = a + bi$ and $z_2 = c + di$, where $a, b, c, d \in \mathbb{R}$.
First, compute the sum $z_1 + z_2$:

$$z_1 + z_2 = (a + bi) + (c + di) = (a + c) + (b + d)i.$$

Now, find the conjugate of $z_1 + z_2$:

$$\overline{z_1 + z_2} = \overline{(a + c) + (b + d)i} = (a + c) - (b + d)i.$$

On the other hand, compute the conjugate of each complex number and then their sum:

$$\overline{z_1} = a - bi, \quad \overline{z_2} = c - di.$$

The sum of the conjugates is:

$$\overline{z_1} + \overline{z_2} = (a - bi) + (c - di) = (a + c) - (b + d)i.$$

We observe that:

$$\overline{z_1 + z_2} = \overline{z_1} + \overline{z_2},$$

which proves the property. ∎

This corollary indicates that conjugation is a linear operation in the set of complex numbers, which facilitates the algebraic analysis of functions and complex expressions.

> (R) The conjugation operation has a geometric interpretation as a reflection over the real axis in the Argand plane. Moreover, the conjugate is useful in defining the magnitude, dividing complex numbers, and solving complex equations.

Exercise 3.6 For each of the following complex numbers, find its conjugate and compute the product of the complex number with its conjugate:
1. $z_1 = 2 + 3i$
2. $z_2 = -4 - 5i$
3. $z_3 = 1 - i$

Then verify that the result equals the square of the magnitude of the complex number.

The concept of the conjugate of a complex number is essential for many advanced operations in algebra and complex analysis. In later chapters, we will see how the conjugate is used to rationalize fractions containing complex numbers in the denominator, and how it simplifies the calculation of limits and the determination of continuity of complex functions. Moreover, the conjugate plays a key role in defining holomorphic functions and in proving fundamental properties in the theory of complex functions.

3.3 De Moivre's Formula and Its Applications

3.3.1 Definition of De Moivre's Formula

Definition 3.3.1 De Moivre's Formula states that for a complex number z in polar form $z = r(\cos\theta + i\sin\theta)$ and an integer n, the following holds:

$$z^n = r^n\left(\cos(n\theta) + i\sin(n\theta)\right)$$

where r is the modulus of the complex number, and θ is its argument. This formula allows for a straightforward computation of powers of complex numbers using trigonometry.

De Moivre's formula is particularly useful for calculating powers and roots of complex numbers and has applications in various fields such as physics and engineering, where periodic signals and oscillations are studied.

■ **Example 3.9** Let us compute $(1+i)^5$ using De Moivre's formula. First, we write the complex number in polar form:

$$r = |1+i| = \sqrt{1^2+1^2} = \sqrt{2}, \quad \theta = \tan^{-1}\left(\frac{1}{1}\right) = \frac{\pi}{4}.$$

Using De Moivre's formula:

$$(1+i)^5 = (\sqrt{2})^5\left(\cos\left(5\cdot\frac{\pi}{4}\right) + i\sin\left(5\cdot\frac{\pi}{4}\right)\right)$$

$$= 4\sqrt{2}\left(\cos\left(\frac{5\pi}{4}\right) + i\sin\left(\frac{5\pi}{4}\right)\right)$$

$$= 4\sqrt{2}\left(-\frac{\sqrt{2}}{2} - i\frac{\sqrt{2}}{2}\right) = -4 - 4i.$$

Thus, $(1+i)^5 = -4 - 4i$. ■

Lema 3.3.1 For a complex number $z = r(\cos\theta + i\sin\theta)$ and an integer n, the magnitude of z^n is given by $|z^n| = |r^n| = r^n$. This means that the modulus of a power of a complex number is simply the power of its original modulus.

This lemma provides a simple way to calculate the modulus of powers of complex numbers without needing to recompute the entire modulus of the resulting expression.

Theorem 3.3.1 De Moivre's formula can also be used to find the nth roots of a complex number. If $z = r(\cos\theta + i\sin\theta)$, then the nth roots are given by:

$$z_k = \sqrt[n]{r}\left(\cos\left(\frac{\theta + 2k\pi}{n}\right) + i\sin\left(\frac{\theta + 2k\pi}{n}\right)\right), \quad k = 0, 1, \ldots, n-1.$$

Demostración. To prove this theorem, let us consider the complex number z in its polar form:

$$z = r(\cos\theta + i\sin\theta).$$

We aim to find the nth roots of the complex number, i.e., we seek complex numbers w such that $w^n = z$.

Assume that w also has a polar form, $w = R(\cos\phi + i\sin\phi)$, where R and ϕ are the modulus and argument of the desired root, respectively. Raising w to the power of n and applying De Moivre's formula, we have:

$$w^n = R^n\left(\cos(n\phi) + i\sin(n\phi)\right).$$

3.3 De Moivre's Formula and Its Applications

We require this to equal the original complex number z:

$$w^n = r(\cos\theta + i\sin\theta).$$

This implies:

$$R^n = r \quad \text{and} \quad n\phi = \theta + 2k\pi, \quad k \in \mathbb{Z}.$$

From $R^n = r$, we find:

$$R = \sqrt[n]{r}.$$

From $n\phi = \theta + 2k\pi$, we solve for ϕ:

$$\phi = \frac{\theta + 2k\pi}{n}, \quad k = 0, 1, \ldots, n-1.$$

Thus, the nth roots of z are given by:

$$z_k = \sqrt[n]{r}\left(\cos\left(\frac{\theta + 2k\pi}{n}\right) + i\sin\left(\frac{\theta + 2k\pi}{n}\right)\right), \quad k = 0, 1, \ldots, n-1.$$

This completes the proof. ∎

This theorem generalizes the use of De Moivre's formula to obtain all the nth roots of a complex number. Each value of k corresponds to a distinct root, and these roots are evenly distributed in the complex plane.

Corollary 3.3.2 The nth roots of a complex number are evenly distributed on a circle of radius $\sqrt[n]{r}$ in the Argand plane, with angles between consecutive roots equal to $\frac{2\pi}{n}$.

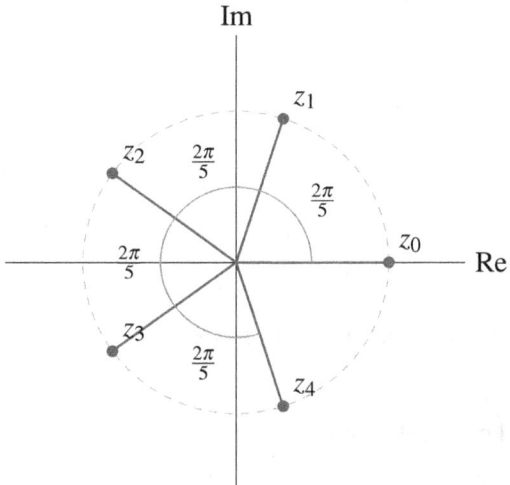

Figura 3.3.1: *Uniform distribution of the nth roots of a complex number in the Argand plane, with equal angles of $\frac{2\pi}{n}$ between them.*

This corollary is useful for visualizing how the roots of a complex number are distributed in the complex plane and shows that all roots have the same magnitude, forming a regular polygon.

> (R) De Moivre's formula is a powerful tool that links trigonometry to complex numbers. It simplifies calculations involving powers and roots and provides an intuitive way to understand the geometric structure of complex numbers in the Argand plane.

Exercise 3.7 Use De Moivre's formula to solve the following problems:
1. Calculate $(\sqrt{3}+i)^4$.
2. Find the cube roots of $z = 8(\cos\pi + i\sin\pi)$.
3. Graph the square roots of $z = 4(\cos\frac{\pi}{3} + i\sin\frac{\pi}{3})$ in the Argand plane.

De Moivre's formula is fundamental for analyzing and manipulating complex numbers in situations requiring powers or roots. In subsequent chapters, we will explore how these tools are applied to problems in complex analysis, such as evaluating complex series and solving differential equations involving complex coefficients. Additionally, the connection between trigonometry and complex numbers facilitated by De Moivre's formula is essential in areas such as physics, wave theory, and signal analysis.

3.3.2 Powers of Complex Numbers Using De Moivre's Formula

Definition 3.3.2 De Moivre's Formula can be used to calculate the powers of a complex number expressed in polar form. If a complex number z is written as $z = r(\cos\theta + i\sin\theta)$, then the nth power of z is given by:

$$z^n = r^n\left(\cos(n\theta) + i\sin(n\theta)\right),$$

where n is an integer. This formula simplifies the computation of powers of complex numbers by utilizing trigonometry.

De Moivre's Formula simplifies raising complex numbers to any integer power by working in terms of their moduli and arguments, streamlining the operation.

■ **Example 3.10** Let us compute $(1+\sqrt{3}i)^6$ using De Moivre's Formula. First, we convert the complex number to its polar form:

$$r = |1+\sqrt{3}i| = \sqrt{1^2 + (\sqrt{3})^2} = \sqrt{1+3} = 2,$$

$$\theta = \tan^{-1}\left(\frac{\sqrt{3}}{1}\right) = \frac{\pi}{3}.$$

Using De Moivre's Formula to compute the sixth power:

$$(1+\sqrt{3}i)^6 = 2^6 \left(\cos\left(6\cdot\frac{\pi}{3}\right) + i\sin\left(6\cdot\frac{\pi}{3}\right)\right),$$

$$= 64\left(\cos 2\pi + i\sin 2\pi\right) = 64(1+0i) = 64.$$

Thus, $(1+\sqrt{3}i)^6 = 64$. ■

Lema 3.3.2 For any complex number $z = r(\cos\theta + i\sin\theta)$ and an integer n, the modulus of z^n is given by $|z^n| = r^n$. This means that when raising a complex number to an integer power, its modulus is raised to the same power.

This lemma shows that the computation of the modulus of a power of a complex number is straightforward and follows a direct rule, similar to multiplying real numbers.

3.3 De Moivre's Formula and Its Applications

Theorem 3.3.3 For two complex numbers $z_1 = r_1(\cos\theta_1 + i\sin\theta_1)$ and $z_2 = r_2(\cos\theta_2 + i\sin\theta_2)$, the nth power of their product $z_1 z_2$ is given by:

$$(z_1 z_2)^n = (r_1 r_2)^n \left(\cos(n(\theta_1 + \theta_2)) + i\sin(n(\theta_1 + \theta_2))\right).$$

Demostración. First, compute the product of the two complex numbers z_1 and z_2:

$$z_1 z_2 = r_1 r_2 \left(\cos(\theta_1 + \theta_2) + i\sin(\theta_1 + \theta_2)\right),$$

using the multiplication property of complex numbers in polar form, where the moduli multiply, and the arguments add.

Now, to calculate the nth power of the product $z_1 z_2$, apply De Moivre's Formula. Using the formula to raise a complex number in polar form to the power n, we have:

$$(z_1 z_2)^n = (r_1 r_2)^n \left(\cos(n(\theta_1 + \theta_2)) + i\sin(n(\theta_1 + \theta_2))\right).$$

This gives the desired result. The proof relies on applying the definition of the product of complex numbers in polar form, followed by the application of De Moivre's Formula. ∎

This theorem extends De Moivre's Formula to the products of complex numbers and enables efficient computation of powers of products by summing their arguments.

Corollary 3.3.4 If z is a complex number with modulus r and argument θ, then:

$$z^{-n} = \frac{1}{r^n} \left(\cos(-n\theta) + i\sin(-n\theta)\right),$$

for any positive integer n. This provides a straightforward way to compute negative powers of complex numbers.

Demostración. A complex number z in polar form is expressed as:

$$z = r(\cos\theta + i\sin\theta).$$

To compute the negative power z^{-n}, we first take the expression for the positive power and invert it:

$$z^{-n} = (r(\cos\theta + i\sin\theta))^{-n}.$$

This simplifies to:

$$z^{-n} = \frac{1}{r^n} \left(\cos(n\theta) + i\sin(n\theta)\right).$$

Using the property of cosine and sine for negative angles:

$$\cos(n\theta) = \cos(-n\theta), \quad \sin(n\theta) = -\sin(-n\theta),$$

we have:

$$z^{-n} = \frac{1}{r^n} \left(\cos(-n\theta) + i\sin(-n\theta)\right).$$

This completes the proof. ∎

This corollary shows that De Moivre's Formula can be applied not only for positive powers but also for computing multiplicative inverses in the set of complex numbers.

 Using De Moivre's Formula to compute powers of complex numbers is particularly useful in solving trigonometric problems and representing periodic signals. The connection between polar form and raising to powers simplifies operations and provides a clear geometric interpretation in the Argand plane.

Exercise 3.8 Use De Moivre's Formula to solve the following problems:
1. Compute $(2-2i)^4$.
2. Find the power $(-1+i\sqrt{3})^3$ and graph the result in the Argand plane.
3. Calculate $(\sqrt{3}+i)^5$ and describe its modulus and argument.

The computation of powers of complex numbers using De Moivre's Formula is a fundamental skill with applications in circuit theory, signal analysis, and other areas of engineering and physics. In subsequent chapters, we will explore how De Moivre's Formula is used to compute roots of complex numbers and how these tools help analyze periodic and oscillatory behaviors. Additionally, we will apply these techniques to solve complex equations and study the algebraic structure of the set of complex numbers.

3.3.3 Nth Roots of a Complex Number

Definition 3.3.3 The **nth roots** of a complex number $z = r(\cos\theta + i\sin\theta)$ are the complex numbers that, when raised to the power n, yield the complex number z. Using the polar form, the n nth roots of z are given by:

$$z_k = \sqrt[n]{r}\left(\cos\left(\frac{\theta + 2k\pi}{n}\right) + i\sin\left(\frac{\theta + 2k\pi}{n}\right)\right), \quad k = 0, 1, \ldots, n-1,$$

where r is the modulus and θ is the argument of the complex number z. The nth roots are distributed uniformly on a circle of radius $\sqrt[n]{r}$ in the complex plane.

This method allows us to find all the nth roots of a complex number and provides a geometric view of how these roots are distributed in the complex plane.

■ **Example 3.11** Let us compute the cube roots of the complex number $z = 8$. First, express z in its polar form:

$$z = 8(\cos 0 + i\sin 0).$$

The three cube roots are given by:

$$z_k = \sqrt[3]{8}\left(\cos\left(\frac{0 + 2k\pi}{3}\right) + i\sin\left(\frac{0 + 2k\pi}{3}\right)\right), \quad k = 0, 1, 2.$$

Calculating each root:

$$z_0 = 2(\cos 0 + i\sin 0) = 2,$$

$$z_1 = 2\left(\cos\frac{2\pi}{3} + i\sin\frac{2\pi}{3}\right) = -1 + i\sqrt{3},$$

$$z_2 = 2\left(\cos\frac{4\pi}{3} + i\sin\frac{4\pi}{3}\right) = -1 - i\sqrt{3}.$$

Thus, the cube roots of $z = 8$ are 2, $-1 + i\sqrt{3}$, and $-1 - i\sqrt{3}$. ■

3.3 De Moivre's Formula and Its Applications

Lema 3.3.3 The nth roots of a complex number z have the same modulus, equal to $\sqrt[n]{r}$, and the angles between consecutive roots are equal to $\frac{2\pi}{n}$. This implies that the roots are symmetrically distributed in the complex plane, forming a regular polygon.

This lemma helps visualize how the nth roots of a complex number are symmetrically located around the origin in the complex plane.

> **Theorem 3.3.5** If $z = r(\cos\theta + i\sin\theta)$ has n nth roots, then these roots are evenly distributed on a circle of radius $\sqrt[n]{r}$ in the complex plane, and the angles between consecutive roots are $\frac{2\pi}{n}$. In particular, the sum of all nth roots of a complex number is equal to zero if $n > 1$.

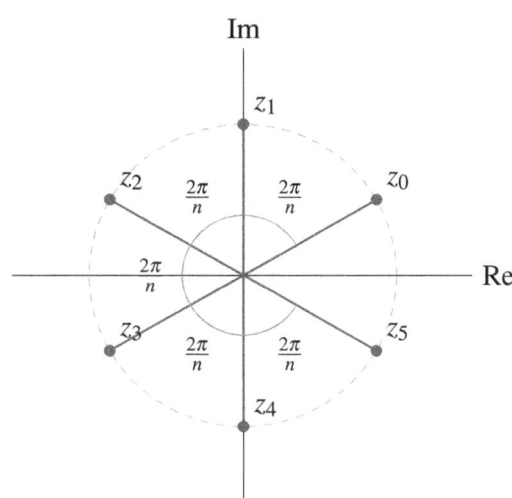

Figura 3.3.2: *Graphical representation of the nth roots of z in the complex plane, uniformly distributed on a circle of radius $\sqrt[n]{r}$ and separated by angles of $\frac{2\pi}{n}$.*

Demostración. Consider a complex number $z = r(\cos\theta + i\sin\theta)$ and its n nth roots, denoted by z_k, where $k = 0, 1, \ldots, n-1$. Using the formula for nth roots of a complex number, we have:

$$z_k = \sqrt[n]{r}\left(\cos\left(\frac{\theta + 2k\pi}{n}\right) + i\sin\left(\frac{\theta + 2k\pi}{n}\right)\right).$$

Observe that the modulus of each root is $\sqrt[n]{r}$, meaning all roots lie on a circle of radius $\sqrt[n]{r}$ in the complex plane.

The angles between consecutive roots are:

$$\Delta\theta = \frac{2\pi}{n},$$

indicating that the roots are evenly spaced around the circle.

To show that the sum of all roots is zero when $n > 1$, sum all the roots:

$$\sum_{k=0}^{n-1} z_k = \sum_{k=0}^{n-1} \sqrt[n]{r}\left(\cos\left(\frac{\theta + 2k\pi}{n}\right) + i\sin\left(\frac{\theta + 2k\pi}{n}\right)\right).$$

Since the angles are evenly distributed, the sum of the real parts and the sum of the imaginary parts cancel out, resulting in:

$$\sum_{k=0}^{n-1} z_k = 0.$$

This completes the proof. ∎

This theorem provides deeper insight into the symmetry and geometric arrangement of the roots in the complex plane.

> **Corollary 3.3.6** The nth roots of a complex number form a regular polygon in the Argand plane, with vertices on a circle of radius $\sqrt[n]{r}$. If $n = 2$, the result is two points symmetric about the real axis, indicating that the square roots of a complex number are symmetric with respect to the real axis.

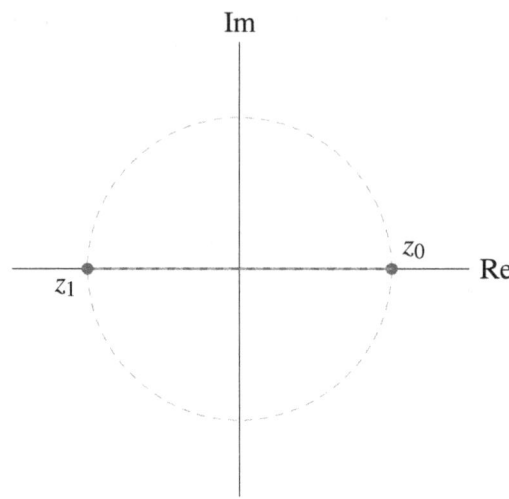

Figura 3.3.3: *The square roots ($n = 2$) of a complex number form two points on a circle of radius $\sqrt[2]{r}$, symmetric about the real axis in the Argand plane.*

This corollary is helpful for understanding the relationship between roots in terms of symmetry and how they can be graphically represented for further analysis.

> (R) Computing the nth roots of a complex number is not only essential in number theory and complex analysis but also has applications in solving differential equations and signal analysis. Roots are used, for instance, to decompose periodic signals into simpler components.

> **Exercise 3.9** Compute the nth roots of the following complex numbers and represent them graphically in the Argand plane:
> 1. Find the cube roots of $z = -8$.
> 2. Determine the four fourth roots of $z = 16(\cos\frac{\pi}{2} + i\sin\frac{\pi}{2})$.
> 3. Calculate the six sixth roots of $z = 64$.
>
> Explain how the roots are distributed in the complex plane and verify that their sum equals zero when $n > 1$.

The nth roots of a complex number play a crucial role in various mathematical and engineering fields. They allow us to solve polynomial equations in the set of complex numbers and understand how these solutions are distributed in the plane. In subsequent chapters, we will explore how root theory connects with differential equations and Fourier transforms, providing powerful tools for decomposing and analyzing periodic phenomena.

3.4 Solved Exercises

3.4 Solved Exercises

Exercise 3.10 Find the cube roots of the complex number $z = 8(\cos 0 + i \sin 0)$ and represent the solutions graphically in the Argand plane.

Demostración. The complex number $z = 8(\cos 0 + i \sin 0)$ is in polar form. To find the cube roots, we use De Moivre's formula for nth roots:

$$z_k = r^{1/n}\left(\cos\left(\frac{\theta + 2k\pi}{n}\right) + i\sin\left(\frac{\theta + 2k\pi}{n}\right)\right)$$

where $r = 8$, $\theta = 0$, $n = 3$, and $k = 0, 1, 2$.

- For $k = 0$:

$$z_0 = 8^{1/3}\left(\cos\left(\frac{0}{3}\right) + i\sin\left(\frac{0}{3}\right)\right) = 2(\cos 0 + i \sin 0) = 2$$

- For $k = 1$:

$$z_1 = 8^{1/3}\left(\cos\left(\frac{2\pi}{3}\right) + i\sin\left(\frac{2\pi}{3}\right)\right) = 2\left(-\frac{1}{2} + i\frac{\sqrt{3}}{2}\right) = -1 + i\sqrt{3}$$

- For $k = 2$:

$$z_2 = 8^{1/3}\left(\cos\left(\frac{4\pi}{3}\right) + i\sin\left(\frac{4\pi}{3}\right)\right) = 2\left(-\frac{1}{2} - i\frac{\sqrt{3}}{2}\right) = -1 - i\sqrt{3}$$

Therefore, the cube roots are 2, $-1 + i\sqrt{3}$, and $-1 - i\sqrt{3}$. These solutions are represented graphically in the Argand plane. ∎

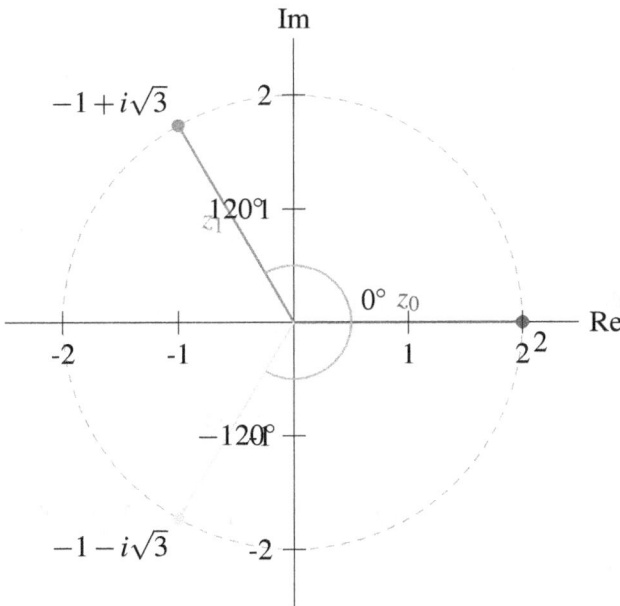

Figura 3.4.1: *Graphical representation of the cube roots of $z = 8$ in the Argand plane.*

Exercise 3.11 Use De Moivre's formula to find $(1+i)^6$ and express the result in rectangular form.

Demostración. First, express $1+i$ in polar form. The modulus is:

$$|1+i| = \sqrt{1^2 + 1^2} = \sqrt{2}$$

The argument is:

$$\arg(1+i) = \tan^{-1}\left(\frac{1}{1}\right) = \frac{\pi}{4}$$

Thus, in polar form:

$$1+i = \sqrt{2}\left(\cos\frac{\pi}{4} + i\sin\frac{\pi}{4}\right)$$

Using De Moivre's formula:

$$(1+i)^6 = \left(\sqrt{2}\right)^6 \left(\cos\left(6 \cdot \frac{\pi}{4}\right) + i\sin\left(6 \cdot \frac{\pi}{4}\right)\right)$$

$$= 8\left(\cos\frac{3\pi}{2} + i\sin\frac{3\pi}{2}\right) = 8(0-i) = -8i$$

Therefore, the result in rectangular form is $-8i$. ∎

Exercise 3.12 Find the conjugate of $z = -3 + 4i$ and calculate the product $z \cdot \bar{z}$. What do you observe about the result?

Demostración. The conjugate of $z = -3 + 4i$ is:

$$\bar{z} = -3 - 4i$$

The product $z \cdot \bar{z}$ is:

$$(-3+4i)(-3-4i) = (-3)^2 - (4i)^2 = 9 - 16i^2$$

Since $i^2 = -1$:

$$= 9 + 16 = 25$$

We observe that the result is a positive real number equal to the square of the modulus of the complex number z:

$$|z|^2 = (-3)^2 + (4)^2 = 9 + 16 = 25$$

Therefore, the product of a complex number and its conjugate always results in a real number equal to the square of the modulus. ∎

3.5 Proposed Exercises

Exercise 3.13 Find the fourth roots of the complex number $z = 16(\cos\frac{\pi}{2} + i\sin\frac{\pi}{2})$ and represent them graphically in the Argand plane.

Demostración. The complex number is in polar form with $r = 16$ and $\theta = \frac{\pi}{2}$. To find the fourth roots, we use De Moivre's formula for nth roots:

$$z_k = r^{1/n}\left(\cos\left(\frac{\theta + 2k\pi}{n}\right) + i\sin\left(\frac{\theta + 2k\pi}{n}\right)\right)$$

where $r = 16$, $\theta = \frac{\pi}{2}$, $n = 4$, and $k = 0, 1, 2, 3$.

- For $k = 0$:
$$z_0 = 16^{1/4}\left(\cos\left(\frac{\pi}{8}\right) + i\sin\left(\frac{\pi}{8}\right)\right) = 2\left(\cos\frac{\pi}{8} + i\sin\frac{\pi}{8}\right)$$

- For $k = 1$:
$$z_1 = 16^{1/4}\left(\cos\left(\frac{\pi}{8} + \frac{\pi}{2}\right) + i\sin\left(\frac{\pi}{8} + \frac{\pi}{2}\right)\right) = 2\left(\cos\frac{5\pi}{8} + i\sin\frac{5\pi}{8}\right)$$

- For $k = 2$:
$$z_2 = 16^{1/4}\left(\cos\left(\frac{\pi}{8} + \pi\right) + i\sin\left(\frac{\pi}{8} + \pi\right)\right) = 2\left(\cos\frac{9\pi}{8} + i\sin\frac{9\pi}{8}\right)$$

- For $k = 3$:
$$z_3 = 16^{1/4}\left(\cos\left(\frac{\pi}{8} + \frac{3\pi}{2}\right) + i\sin\left(\frac{\pi}{8} + \frac{3\pi}{2}\right)\right) = 2\left(\cos\frac{13\pi}{8} + i\sin\frac{13\pi}{8}\right)$$

Thus, the four fourth roots are $2(\cos\frac{\pi}{8} + i\sin\frac{\pi}{8})$, $2(\cos\frac{5\pi}{8} + i\sin\frac{5\pi}{8})$, $2(\cos\frac{9\pi}{8} + i\sin\frac{9\pi}{8})$, and $2(\cos\frac{13\pi}{8} + i\sin\frac{13\pi}{8})$. These solutions are represented graphically in the Argand plane. ∎

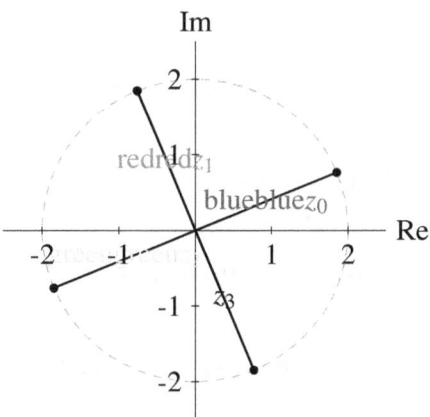

Figura 3.4.2: *Graphical representation of the fourth roots of $z = 16\left(\cos\frac{\pi}{2} + i\sin\frac{\pi}{2}\right)$ in the Argand plane.*

3.5 Proposed Exercises

3.5.1 Representation of Complex Numbers in the Plane

Exercise 3.14 Graph the complex numbers $z_1 = 3+4i$ and $z_2 = -2+i$ in the Argand plane.

Exercise 3.15 Find the modulus and argument of the complex number $z = -1+i\sqrt{3}$.

Exercise 3.16 Describe geometrically the multiplication of the complex number $z = 2+2i$ by i in the complex plane.

Exercise 3.17 Determine the conjugate of the complex number $z = 5-3i$ and graph both in the Argand plane.

Exercise 3.18 Find the modulus of $z_1 z_2$ if $z_1 = 3+i$ and $z_2 = 1-2i$.

3.5.2 Operations with Complex Numbers

Exercise 3.19 Compute $(2+3i)+(-1+5i)$ and graph the result in the complex plane.

Exercise 3.20 Solve the expression $(3-i)-(1+2i)$ and find the modulus of the result.

Exercise 3.21 Prove that the product of a complex number by its conjugate is a real number.

Exercise 3.22 Find the additive and multiplicative inverses of the complex number $z = 4+i$.

Exercise 3.23 Compute $(1+i)^3$ using De Moivre's formula.

3.5.3 De Moivre's Formula and Applications

Exercise 3.24 Use De Moivre's formula to compute $(\sqrt{3}+i)^4$.

Exercise 3.25 Find all cube roots of the complex number $z = 8(\cos \pi + i \sin \pi)$.

Exercise 3.26 Find the modulus and argument of $(1-i)^5$ using De Moivre's formula.

Exercise 3.27 Graph the fourth roots of $z = 16$ in the Argand plane.

Exercise 3.28 Prove that the sixth roots of 1 are evenly distributed in the complex plane.

4. Vectors in $\mathbb{R}^2$ and $\mathbb{R}^3$

4.1 Definition and Operations with Vectors

4.1.1 Definition of Vectors

Definition 4.1.1 A **vector** in $\mathbb{R}^2$ is an ordered pair of real numbers (x,y), while a vector in $\mathbb{R}^3$ is an ordered triple of real numbers (x,y,z). A vector can be graphically represented as an arrow indicating a direction and a magnitude in space. The elements x,y,z are called the **components** of the vector.

■ **Example 4.1** The vector $\vec{v} = (3,4)$ in $\mathbb{R}^2$ represents a magnitude of 5 units (calculated as $\sqrt{3^2 + 4^2} = 5$) and a direction that can be visualized in the Cartesian plane. ■

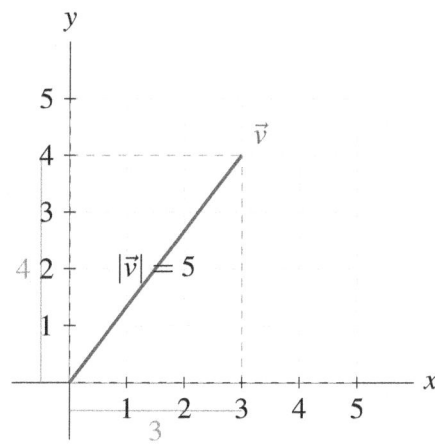

Figura 4.1.1: *Representation of the vector $\vec{v} = (3,4)$ in the Cartesian plane, showing its components and magnitude.*

R It is important to note that vectors can be used to represent displacements, forces, and velocities in the plane or in three-dimensional space. This makes vectors fundamental tools in physics and analytic geometry.

Lema 4.1.1 Given a vector $\vec{v} = (x, y, z)$ in $\mathbb{R}^3$, its norm or magnitude is given by:

$$\|\vec{v}\| = \sqrt{x^2 + y^2 + z^2}$$

This property of vectors allows us to measure their "size" in space.

Theorem 4.1.1 If two vectors $\vec{u}$ and $\vec{v}$ in $\mathbb{R}^2$ or $\mathbb{R}^3$ have the same magnitude and direction, then the vectors are equal. Formally, if $\vec{u} = (x_1, y_1, z_1)$ and $\vec{v} = (x_2, y_2, z_2)$, then:

$$\vec{u} = \vec{v} \iff x_1 = x_2, y_1 = y_2, z_1 = z_2$$

Demostración. Assume that vectors $\vec{u}$ and $\vec{v}$ have the same magnitude and direction. This means that, for $\vec{u} = (x_1, y_1, z_1)$ and $\vec{v} = (x_2, y_2, z_2)$:
1. The magnitude of $\vec{u}$ equals the magnitude of $\vec{v}$:

$$\|\vec{u}\| = \|\vec{v}\} \implies \sqrt{x_1^2 + y_1^2 + z_1^2} = \sqrt{x_2^2 + y_2^2 + z_2^2}.$$

2. The direction of $\vec{u}$ and $\vec{v}$ is also the same. This implies that each corresponding component of the vectors must be equal, as the only way for two vectors to have the same direction and magnitude is if their components are equal. Therefore:

$$x_1 = x_2, \quad y_1 = y_2, \quad z_1 = z_2.$$

Thus, we conclude:

$$\vec{u} = \vec{v} \iff x_1 = x_2, \quad y_1 = y_2, \quad z_1 = z_2.$$

This completes the proof. ■

Corollary 4.1.2 Two vectors $\vec{u}$ and $\vec{v}$ in $\mathbb{R}^2$ are perpendicular if and only if their **dot product** is zero:

$$\vec{u} \cdot \vec{v} = 0$$

This condition can be used to verify the orthogonality of two vectors in the plane.

■ **Example 4.2** Consider the vector $\vec{w} = (-2, 4, 1)$ in $\mathbb{R}^3$. Its norm is:

$$\|\vec{w}\| = \sqrt{(-2)^2 + 4^2 + 1^2} = \sqrt{21}$$

■

As we can see from the previous examples and lemmas, vectors in $\mathbb{R}^2$ and $\mathbb{R}^3$ have a wide range of properties that allow operations such as addition, subtraction, and magnitude calculation, as well as determining geometric relationships like orthogonality. These properties are fundamental in the study of analytic geometry and in solving physical and engineering problems.

4.1.2 Addition and Subtraction of Vectors

4.1 Definition and Operations with Vectors

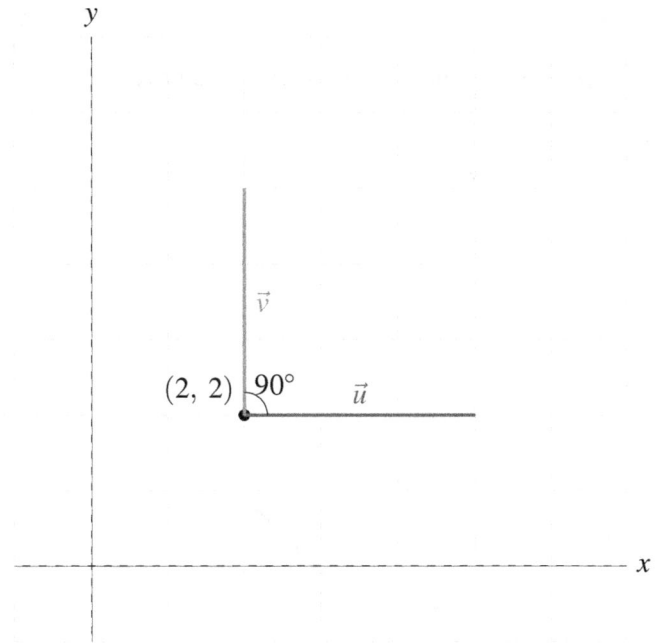

Figura 4.1.2: *Two perpendicular vectors $\vec{u}$ and $\vec{v}$ in $\mathbb{R}^2$, originating from the point $(2,2)$, showing that $\vec{u} \cdot \vec{v} = 0$.*

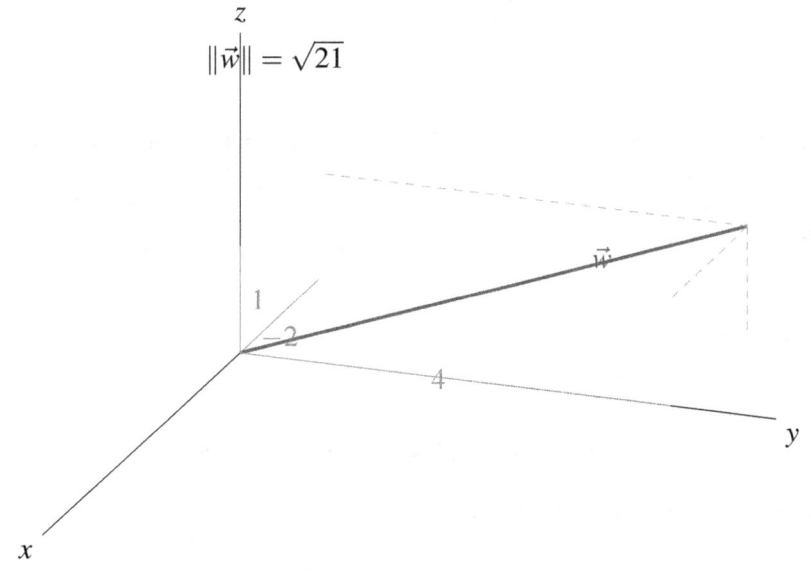

Figura 4.1.3: *Representation of the vector $\vec{w} = (-2, 4, 1)$ in $\mathbb{R}^3$, showing its norm $\|\vec{w}\| = \sqrt{21}$.*

Definition 4.1.2 **Vector addition** is an operation performed component-wise. If we have two vectors in $\mathbb{R}^2$, $\vec{u} = (u_1, u_2)$ and $\vec{v} = (v_1, v_2)$, the sum is defined as:

$$\vec{u} + \vec{v} = (u_1 + v_1, u_2 + v_2)$$

Similarly, in $\mathbb{R}^3$, for $\vec{u} = (u_1, u_2, u_3)$ and $\vec{v} = (v_1, v_2, v_3)$, we have:

$$\vec{u} + \vec{v} = (u_1 + v_1, u_2 + v_2, u_3 + v_3)$$

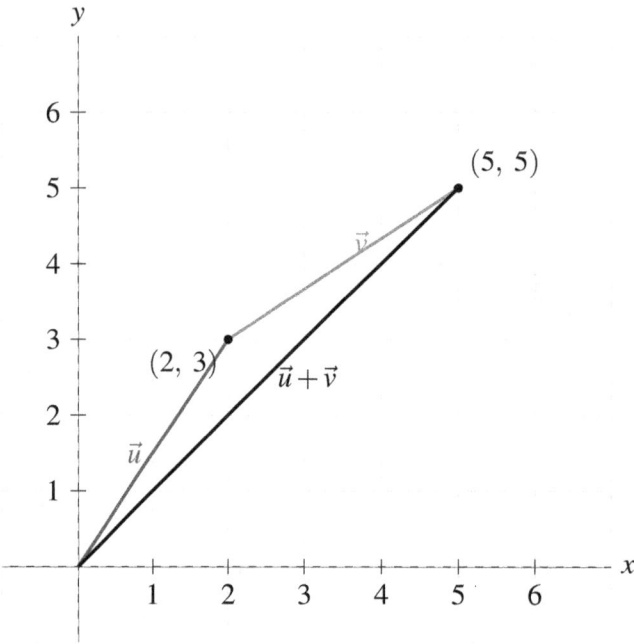

Figura 4.1.4: *Addition of consecutive vectors: $\vec{u}$ from $(0,0)$ to $(2,3)$, $\vec{v}$ from $(2,3)$ to $(5,5)$, and the resultant vector $\vec{u} + \vec{v}$ from $(0,0)$ to $(5,5)$.*

R   Vector addition is commutative and associative, meaning that the order of the vectors does not affect the result: $\vec{u} + \vec{v} = \vec{v} + \vec{u}$ and $(\vec{u} + \vec{v}) + \vec{w} = \vec{u} + (\vec{v} + \vec{w})$.

Lema 4.1.2 The subtraction of two vectors $\vec{u}$ and $\vec{v}$ is defined as the addition of $\vec{u}$ with the opposite of $\vec{v}$:

$$\vec{u} - \vec{v} = \vec{u} + (-\vec{v})$$

where $-\vec{v} = (-v_1, -v_2, -v_3)$ in $\mathbb{R}^3$.

■ **Example 4.3** If $\vec{a} = (3, 4)$ and $\vec{b} = (1, -2)$, the sum of the vectors is:

$$\vec{a} + \vec{b} = (3 + 1, 4 + (-2)) = (4, 2)$$

While the subtraction is:

$$\vec{a} - \vec{b} = (3 - 1, 4 - (-2)) = (2, 6)$$

■

4.1 Definition and Operations with Vectors

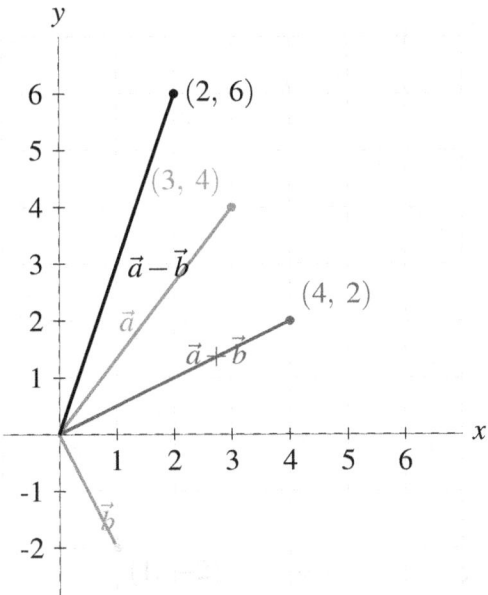

Figura 4.1.5: *Graph of vectors $\vec{a}$, $\vec{b}$, $\vec{a}+\vec{b}$, and $\vec{a}-\vec{b}$ in $\mathbb{R}^2$.*

Theorem 4.1.3 For any vector $\vec{u}$ in $\mathbb{R}^n$, adding the zero vector $\vec{0}$ does not change the original vector:

$$\vec{u} + \vec{0} = \vec{u}$$

where $\vec{0} = (0, 0, \ldots, 0)$ is the zero vector of dimension n.

Corollary 4.1.4 For any vector $\vec{v}$ in $\mathbb{R}^n$, the following holds:

$$\vec{v} + (-\vec{v}) = \vec{0}$$

This result shows that adding a vector to its opposite results in the zero vector.

Demostración. By definition, the vector $-\vec{v}$ has the same magnitude as $\vec{v}$ but in the opposite direction. When we add $\vec{v}$ and $-\vec{v}$, each component cancels out:

$$\vec{v} + (-\vec{v}) = (v_1, v_2, \ldots, v_n) + (-v_1, -v_2, \ldots, -v_n) = (0, 0, \ldots, 0) = \vec{0}.$$

Thus, the sum of a vector with its opposite always results in the zero vector $\vec{0}$. ∎

Exercise 4.1 Given $\vec{u} = (2, -3, 4)$ and $\vec{v} = (-1, 5, 0)$ in $\mathbb{R}^3$, calculate the sum $\vec{u} + \vec{v}$.

Exercise 4.2 Find the difference of the vectors $\vec{p} = (1, -1, 3)$ and $\vec{q} = (4, 2, -1)$ in $\mathbb{R}^3$.

■ **Example 4.4** Consider the vectors $\vec{x} = (1, 0, -1)$ and $\vec{y} = (2, 3, 1)$ in $\mathbb{R}^3$. Then:

$$\vec{x} + \vec{y} = (1+2, 0+3, -1+1) = (3, 3, 0)$$

and the difference is:

$$\vec{x} - \vec{y} = (1-2, 0-3, -1-1) = (-1, -3, -2)$$

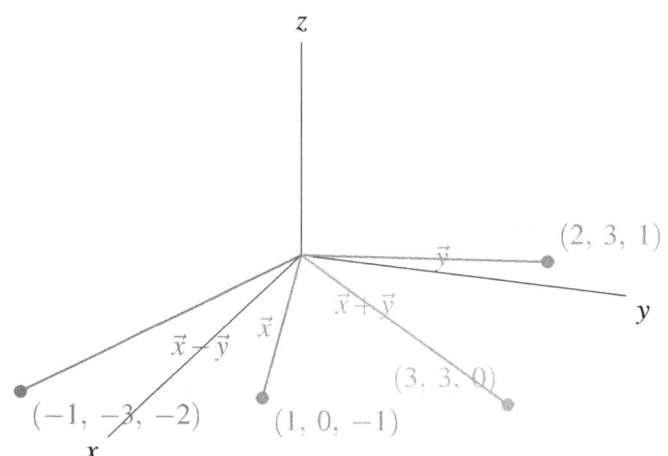

Figura 4.1.6: *Graphical representation of $\vec{x}$, $\vec{y}$, $\vec{x}+\vec{y}$, and $\vec{x}-\vec{y}$ in $\mathbb{R}^3$.*

In summary, vector addition and subtraction are fundamental operations that allow combining magnitudes and directions. These operations are essential for working with vectors in geometry, physics, and other fields of applied mathematics.

4.1.3 Scalar Multiplication

Definition 4.1.3 The **scalar multiplication of a vector** is the operation in which each component of the vector is multiplied by a real number. Formally, if $\vec{v} = (v_1, v_2, v_3)$ is a vector in $\mathbb{R}^3$ and $k \in \mathbb{R}$ is a scalar, then the product $k\vec{v}$ is defined as:

$$k\vec{v} = (kv_1, kv_2, kv_3)$$

■ **Example 4.5** Consider the vector $\vec{v} = (2, -3, 4)$ and the scalar $k = -2$. The scalar multiplication of $\vec{v}$ by k is:

$$-2\vec{v} = (-2 \cdot 2, -2 \cdot (-3), -2 \cdot 4) = (-4, 6, -8)$$

Lema 4.1.3 For any scalar $k \in \mathbb{R}$ and any vector $\vec{v} \in \mathbb{R}^n$, the following holds:

$$k(0\vec{v}) = 0 \quad \text{and} \quad 0\vec{v} = \vec{0}$$

where $\vec{0}$ is the zero vector.

Theorem 4.1.5 Let $k \in \mathbb{R}$ and $\vec{u}, \vec{v} \in \mathbb{R}^n$. The following distributive property holds:

$$k(\vec{u} + \vec{v}) = k\vec{u} + k\vec{v}$$

This property shows how the scalar affects each component of the vector sum.

Demostración. Suppose the vectors $\vec{u}$ and $\vec{v}$ are given by:

$$\vec{u} = (u_1, u_2, \ldots, u_n), \quad \vec{v} = (v_1, v_2, \ldots, v_n)$$

4.1 Definition and Operations with Vectors

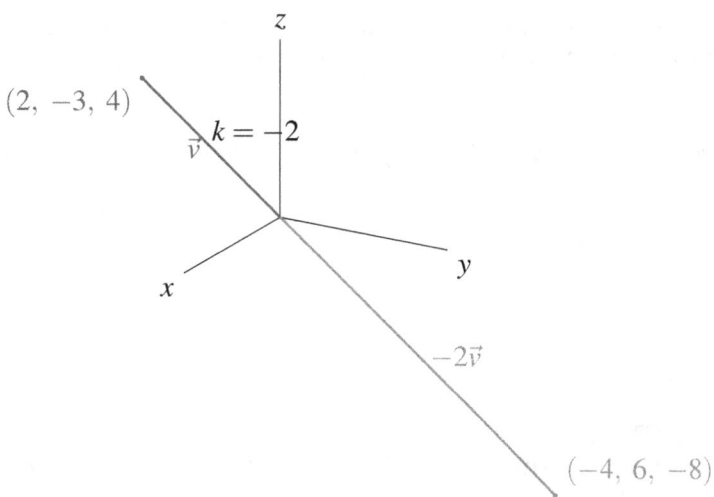

Figura 4.1.7: *Representation of the vector $\vec{v}$ and its scalar multiple $-2\vec{v}$ in $\mathbb{R}^3$.*

Then the sum of the vectors is:

$$\vec{u} + \vec{v} = (u_1 + v_1, u_2 + v_2, \ldots, u_n + v_n)$$

Multiplying by the scalar k:

$$k(\vec{u} + \vec{v}) = k(u_1 + v_1, u_2 + v_2, \ldots, u_n + v_n) = (ku_1 + kv_1, ku_2 + kv_2, \ldots, ku_n + kv_n)$$

This can be rewritten as:

$$k\vec{u} + k\vec{v} = (ku_1, ku_2, \ldots, ku_n) + (kv_1, kv_2, \ldots, kv_n) = (ku_1 + kv_1, ku_2 + kv_2, \ldots, ku_n + kv_n)$$

Thus:

$$k(\vec{u} + \vec{v}) = k\vec{u} + k\vec{v}$$

This proves the distributive property of scalar multiplication with vector addition. ∎

Corollary 4.1.6 For two scalars $k_1, k_2 \in \mathbb{R}$ and a vector $\vec{v} \in \mathbb{R}^n$, the following holds:

$$(k_1 + k_2)\vec{v} = k_1\vec{v} + k_2\vec{v}$$

Demostración. Suppose the vector $\vec{v}$ is given by:

$$\vec{v} = (v_1, v_2, \ldots, v_n)$$

Then, multiplying $\vec{v}$ by the scalar $k_1 + k_2$:

$$(k_1 + k_2)\vec{v} = ((k_1 + k_2)v_1, (k_1 + k_2)v_2, \ldots, (k_1 + k_2)v_n)$$

Using the distributive property of multiplication, we can write:

$$(k_1 + k_2)\vec{v} = (k_1 v_1 + k_2 v_1, k_1 v_2 + k_2 v_2, \ldots, k_1 v_n + k_2 v_n)$$

This can be rewritten as the sum of two vectors:

$$k_1\vec{v} + k_2\vec{v} = (k_1 v_1, k_1 v_2, \ldots, k_1 v_n) + (k_2 v_1, k_2 v_2, \ldots, k_2 v_n)$$

Thus:

$$(k_1 + k_2)\vec{v} = k_1\vec{v} + k_2\vec{v}$$

∎

(R) The scalar multiplication of a vector does not change the direction of the vector unless the scalar is negative, in which case the direction is reversed.

Exercise 4.3 Given the vector $\vec{a} = (3, -5, 1)$ and the scalar $k = 4$, calculate the product $k\vec{a}$.

Exercise 4.4 If $\vec{b} = (-2, 6, 0)$ and the scalar is $k = -3$, determine $k\vec{b}$ and describe the direction of the resulting vector.

4.2 Dot Product and Cross Product

4.2.1 Definition of the Dot Product

Definition 4.2.1 The **dot product** (or **scalar product**) of two vectors $\vec{u} = (u_1, u_2, u_3)$ and $\vec{v} = (v_1, v_2, v_3)$ in $\mathbb{R}^3$ is defined as:

$$\vec{u} \cdot \vec{v} = u_1 v_1 + u_2 v_2 + u_3 v_3$$

The result of the dot product is a real number. This operation is fundamental in analytic geometry, with applications such as determining the angle between two vectors and verifying orthogonality.

■ **Example 4.6** Consider the vectors $\vec{u} = (1, 2, 3)$ and $\vec{v} = (4, -5, 6)$. The dot product is:

$$\vec{u} \cdot \vec{v} = (1)(4) + (2)(-5) + (3)(6) = 4 - 10 + 18 = 12$$

Lema 4.2.1 The dot product of a vector with itself equals the square of its magnitude. For any vector $\vec{u} \in \mathbb{R}^3$:

$$\vec{u} \cdot \vec{u} = \|\vec{u}\|^2$$

where $\|\vec{u}\|$ is the norm (length) of $\vec{u}$.

Demostración. Let $\vec{u} = (u_1, u_2, u_3)$ be a vector in $\mathbb{R}^3$. The dot product of $\vec{u}$ with itself is defined as:

$$\vec{u} \cdot \vec{u} = u_1^2 + u_2^2 + u_3^2$$

On the other hand, the norm of $\vec{u}$ is defined as:

$$\|\vec{u}\| = \sqrt{u_1^2 + u_2^2 + u_3^2}$$

Squaring the norm gives:

$$\|\vec{u}\|^2 = u_1^2 + u_2^2 + u_3^2$$

Thus:

$$\vec{u} \cdot \vec{u} = \|\vec{u}\|^2$$

4.2 Dot Product and Cross Product

Theorem 4.2.1 Let $\vec{u}, \vec{v} \in \mathbb{R}^3$. The angle θ between the vectors $\vec{u}$ and $\vec{v}$ is related to the dot product by the formula:

$$\vec{u} \cdot \vec{v} = \|\vec{u}\| \|\vec{v}\| \cos(\theta)$$

This expression is useful for calculating the angle between two vectors or determining if they are orthogonal (when the dot product equals zero).

Demostración. The formula for the dot product is derived from its geometric definition. Let $\vec{u}$ and $\vec{v}$ be two vectors in $\mathbb{R}^3$. If we represent $\vec{u}$ and $\vec{v}$ as arrows originating from the same point, the dot product can be expressed in terms of the magnitudes of the vectors and the angle θ between them. By definition, the dot product $\vec{u} \cdot \vec{v}$ can be computed as the sum of the products of their components:

$$\vec{u} \cdot \vec{v} = u_1 v_1 + u_2 v_2 + u_3 v_3$$

Simultaneously, the dot product can also be described as:

$$\vec{u} \cdot \vec{v} = \|\vec{u}\| \|\vec{v}\| \cos(\theta)$$

where $\|\vec{u}\|$ and $\|\vec{v}\|$ are the magnitudes of the vectors $\vec{u}$ and $\vec{v}$, respectively, and θ is the angle between the vectors. This formula reflects the geometric nature of the dot product, considering both vector magnitudes and their angular relationship.
If $\theta = 90°$, then $\cos(\theta) = 0$, and the dot product $\vec{u} \cdot \vec{v} = 0$, implying that the vectors are orthogonal. Thus, the formula for the dot product in terms of the angle is demonstrated. ∎

Corollary 4.2.2 Two vectors $\vec{u}$ and $\vec{v}$ in $\mathbb{R}^3$ are **orthogonal** if and only if their dot product equals zero:

$$\vec{u} \cdot \vec{v} = 0$$

Demostración. By definition, the dot product of two vectors $\vec{u}$ and $\vec{v}$ in $\mathbb{R}^3$ is given by:

$$\vec{u} \cdot \vec{v} = \|\vec{u}\| \|\vec{v}\| \cos(\theta)$$

where $\|\vec{u}\|$ and $\|\vec{v}\|$ are the magnitudes of the vectors, and θ is the angle between them.
If the vectors $\vec{u}$ and $\vec{v}$ are orthogonal, the angle between them is $\theta = 90°$, and we know that $\cos(90°) = 0$. Thus:

$$\vec{u} \cdot \vec{v} = \|\vec{u}\| \|\vec{v}\| \cos(90°) = 0$$

Conversely, if the dot product $\vec{u} \cdot \vec{v} = 0$, then $\cos(\theta) = 0$, which implies that $\theta = 90°$, and therefore, the vectors are orthogonal.
This proves that two vectors are orthogonal if and only if their dot product equals zero. ∎

> (R) The dot product is commutative, meaning $\vec{u} \cdot \vec{v} = \vec{v} \cdot \vec{u}$. This property is fundamental and used in various proofs and applications.

> **Exercise 4.5** Given the vector $\vec{a} = (2, -1, 3)$ and the vector $\vec{b} = (-1, 4, 2)$, calculate the dot product $\vec{a} \cdot \vec{b}$.

> **Exercise 4.6** Determine whether the vectors $\vec{u} = (1, 0, -1)$ and $\vec{v} = (0, 1, 1)$ are orthogonal by calculating their dot product.

4.2.2 Properties of the Dot Product

> **Definition 4.2.2** The **dot product** is an operation that associates two vectors $\vec{u}, \vec{v} \in \mathbb{R}^n$ with a real number. In addition to the definition of the dot product, several properties help to better understand how vectors interact with each other. These include:
> - **Commutative:** $\vec{u} \cdot \vec{v} = \vec{v} \cdot \vec{u}$
> - **Distributive:** $\vec{u} \cdot (\vec{v} + \vec{w}) = \vec{u} \cdot \vec{v} + \vec{u} \cdot \vec{w}$
> - **Scaling:** $(k\vec{u}) \cdot \vec{v} = k(\vec{u} \cdot \vec{v})$, where $k \in \mathbb{R}$ is a scalar.

■ **Example 4.7** Let $\vec{a} = (2, 3)$ and $\vec{b} = (-1, 4)$. We verify the commutative property of the dot product:

$$\vec{a} \cdot \vec{b} = (2)(-1) + (3)(4) = -2 + 12 = 10$$

$$\vec{b} \cdot \vec{a} = (-1)(2) + (4)(3) = -2 + 12 = 10$$

Thus, $\vec{a} \cdot \vec{b} = \vec{b} \cdot \vec{a}$. ■

Lema 4.2.2 For any vector $\vec{u} \in \mathbb{R}^n$, the dot product of $\vec{u}$ with itself is non-negative:

$$\vec{u} \cdot \vec{u} \geq 0$$

and is equal to zero if and only if $\vec{u} = \vec{0}$.

Demostración. By definition, the dot product of a vector $\vec{u} = (u_1, u_2, \ldots, u_n)$ with itself is computed as:

$$\vec{u} \cdot \vec{u} = u_1^2 + u_2^2 + \cdots + u_n^2$$

Since each term $u_i^2 \geq 0$ for any real number u_i, the total value of $\vec{u} \cdot \vec{u}$ is also non-negative:

$$\vec{u} \cdot \vec{u} = u_1^2 + u_2^2 + \cdots + u_n^2 \geq 0$$

Moreover, $\vec{u} \cdot \vec{u} = 0$ if and only if all components of the vector are zero, i.e., $\vec{u} = \vec{0}$.
This proves that the dot product of a vector with itself is always non-negative and equals zero only when the vector is the zero vector. ∎

> **Theorem 4.2.3** The dot product satisfies the distributive property with respect to vector addition. For $\vec{u}, \vec{v}, \vec{w} \in \mathbb{R}^n$:
>
> $$\vec{u} \cdot (\vec{v} + \vec{w}) = \vec{u} \cdot \vec{v} + \vec{u} \cdot \vec{w}$$
>
> This property is fundamental in linear algebra, as it allows the dot product of a vector with a linear combination of other vectors to be expressed in a simplified way.

4.2 Dot Product and Cross Product

Demostración. Let $\vec{u} = (u_1, u_2, \ldots, u_n)$, $\vec{v} = (v_1, v_2, \ldots, v_n)$, and $\vec{w} = (w_1, w_2, \ldots, w_n)$ in $\mathbb{R}^n$.
The vector $\vec{v} + \vec{w}$ is defined as:

$$\vec{v} + \vec{w} = (v_1 + w_1, v_2 + w_2, \ldots, v_n + w_n)$$

The dot product of $\vec{u}$ with $\vec{v} + \vec{w}$ is calculated as:

$$\vec{u} \cdot (\vec{v} + \vec{w}) = u_1(v_1 + w_1) + u_2(v_2 + w_2) + \cdots + u_n(v_n + w_n)$$

Applying the distributive property of real numbers:

$$\vec{u} \cdot (\vec{v} + \vec{w}) = (u_1 v_1 + u_1 w_1) + (u_2 v_2 + u_2 w_2) + \cdots + (u_n v_n + u_n w_n)$$

Reorganizing terms, we get:

$$\vec{u} \cdot (\vec{v} + \vec{w}) = (u_1 v_1 + u_2 v_2 + \cdots + u_n v_n) + (u_1 w_1 + u_2 w_2 + \cdots + u_n w_n)$$

This equals:

$$\vec{u} \cdot \vec{v} + \vec{u} \cdot \vec{w}$$

Thus, we have shown that:

$$\vec{u} \cdot (\vec{v} + \vec{w}) = \vec{u} \cdot \vec{v} + \vec{u} \cdot \vec{w}$$

∎

Corollary 4.2.4 For any scalar $k \in \mathbb{R}$ and vectors $\vec{u}, \vec{v} \in \mathbb{R}^n$:

$$(k\vec{u}) \cdot \vec{v} = k(\vec{u} \cdot \vec{v})$$

This implies that the dot product is linear with respect to scalar multiplication.

Demostración. Let $k \in \mathbb{R}$ and $\vec{u} = (u_1, u_2, \ldots, u_n)$, $\vec{v} = (v_1, v_2, \ldots, v_n)$ in $\mathbb{R}^n$.
The dot product of $k\vec{u}$ with $\vec{v}$ is calculated as:

$$(k\vec{u}) \cdot \vec{v} = (ku_1, ku_2, \ldots, ku_n) \cdot (v_1, v_2, \ldots, v_n)$$

Using the definition of the dot product, we have:

$$(k\vec{u}) \cdot \vec{v} = ku_1 v_1 + ku_2 v_2 + \cdots + ku_n v_n$$

Factoring k:

$$(k\vec{u}) \cdot \vec{v} = k(u_1 v_1 + u_2 v_2 + \cdots + u_n v_n)$$

Thus:

$$(k\vec{u}) \cdot \vec{v} = k(\vec{u} \cdot \vec{v})$$

∎

R The dot product plays a key role in determining the orthogonality of vectors. If $\vec{u} \cdot \vec{v} = 0$, the vectors are perpendicular to each other.

Exercise 4.7 Verify the distributive property for the vectors $\vec{a} = (1,2,3)$, $\vec{b} = (4,-1,0)$, and $\vec{c} = (-2,3,1)$ by calculating $\vec{a} \cdot (\vec{b}+\vec{c})$ and comparing it with $\vec{a} \cdot \vec{b} + \vec{a} \cdot \vec{c}$.

Exercise 4.8 Determine whether the vectors $\vec{u} = (3,-2)$ and $\vec{v} = (-2,-3)$ are orthogonal by calculating their dot product.

4.2.3 Definition of the Cross Product

Definition 4.2.3 The **cross product** of two vectors $\vec{u} = (u_1, u_2, u_3)$ and $\vec{v} = (v_1, v_2, v_3)$ in $\mathbb{R}^3$ is defined as:

$$\vec{u} \times \vec{v} = (u_2 v_3 - u_3 v_2, u_3 v_1 - u_1 v_3, u_1 v_2 - u_2 v_1)$$

The result is a vector perpendicular to both $\vec{u}$ and $\vec{v}$, which is essential for determining orthogonality in three-dimensional space.

■ **Example 4.8** Let us compute the cross product of the vectors $\vec{a} = (1,0,2)$ and $\vec{b} = (3,-1,4)$:

$$\vec{a} \times \vec{b} = (0 \cdot 4 - 2 \cdot (-1), 2 \cdot 3 - 1 \cdot 4, 1 \cdot (-1) - 0 \cdot 3) = (2,2,-1)$$

Thus, $\vec{a} \times \vec{b} = (2,2,-1)$.

■

Lema 4.2.3 The cross product of two parallel vectors is the zero vector. That is, if $\vec{u} = k\vec{v}$ for some scalar k, then $\vec{u} \times \vec{v} = \vec{0}$.

Demostración. Let $\vec{u} = k\vec{v}$, where $k \in \mathbb{R}$. Recall the definition of the cross product in $\mathbb{R}^3$:

$$\vec{u} \times \vec{v} = \|\vec{u}\| \|\vec{v}\| \sin(\theta) \hat{n}$$

where θ is the angle between $\vec{u}$ and $\vec{v}$, and $\hat{n}$ is a unit vector perpendicular to both. Since $\vec{u}$ and $\vec{v}$ are parallel, $\theta = 0$ or $\theta = \pi$. In both cases, $\sin(\theta) = 0$. Therefore:

$$\vec{u} \times \vec{v} = \|\vec{u}\| \|\vec{v}\| \sin(\theta) \hat{n} = 0$$

Thus, the cross product of two parallel vectors is the zero vector:

$$\vec{u} \times \vec{v} = \vec{0}$$

∎

4.2 Dot Product and Cross Product

Theorem 4.2.5 The magnitude of the cross product of two vectors $\vec{u}$ and $\vec{v}$ is given by:

$$|\vec{u} \times \vec{v}| = |\vec{u}||\vec{v}|\sin\theta$$

where θ is the angle between the vectors $\vec{u}$ and $\vec{v}$. This formula is useful for determining the area of the parallelogram formed by the vectors.

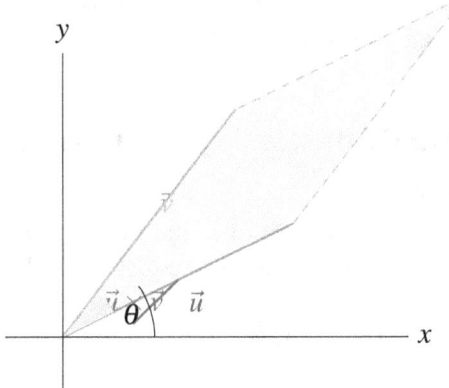

Figura 4.2.1: *Graphical representation of vectors $\vec{u}$ and $\vec{v}$, the angle θ between them, and the parallelogram formed, with an area of $|\vec{u} \times \vec{v}| = |\vec{u}||\vec{v}|\sin\theta$.*

Demostración. Let $\vec{u}$ and $\vec{v}$ be two vectors in $\mathbb{R}^3$, and let θ be the angle between them. By definition, the cross product $\vec{u} \times \vec{v}$ is a vector perpendicular to both $\vec{u}$ and $\vec{v}$, and its magnitude is given by:

$$|\vec{u} \times \vec{v}| = |\vec{u}||\vec{v}|\sin(\theta)$$

where $|\vec{u}|$ and $|\vec{v}|$ are the magnitudes of the vectors $\vec{u}$ and $\vec{v}$, respectively, and θ is the angle between them.

The magnitude of the cross product $|\vec{u} \times \vec{v}|$ corresponds to the area of the parallelogram formed by the vectors $\vec{u}$ and $\vec{v}$. This is because the area of a parallelogram is equal to the base multiplied by the height, and $|\vec{u}||\vec{v}|\sin(\theta)$ represents precisely that geometric relationship.

$$|\vec{u} \times \vec{v}| = |\vec{u}||\vec{v}|\sin(\theta)$$

Thus, we have proven the formula for the magnitude of the cross product.

∎

Corollary 4.2.6 If the cross product of two vectors is not zero ($\vec{u} \times \vec{v} \neq \vec{0}$), then the vectors $\vec{u}$ and $\vec{v}$ are not parallel and, therefore, are linearly independent.

Demostración. If the cross product of two vectors $\vec{u}$ and $\vec{v}$ is not equal to the zero vector ($\vec{u} \times \vec{v} \neq \vec{0}$), this implies that $\sin(\theta) \neq 0$, where θ is the angle between the vectors.

$\sin(\theta) \neq 0$ means that the angle θ is neither $0°$ nor $180°$, i.e., the vectors are not parallel. If the vectors are not parallel, they are not scalar multiples of each other and, therefore, are linearly independent.

∎

> The cross product is useful in physics for describing quantities such as torque and magnetic force, as it generates a vector orthogonal to the plane defined by the original vectors.

Exercise 4.9 Calculate the cross product of the vectors $\vec{u} = (2,1,3)$ and $\vec{v} = (-1,4,0)$. Verify whether the resulting vector is orthogonal to both $\vec{u}$ and $\vec{v}$.

Exercise 4.10 Determine whether the cross product of the vectors $\vec{a} = (1,2,3)$ and $\vec{b} = (2,4,6)$ equals the zero vector. What can you conclude about the relationship between $\vec{a}$ and $\vec{b}$?

4.3 Parametric Equations and Cartesian Equations of Lines and Planes

4.3.1 Parametric Equations of a Line

Definition 4.3.1 The **parametric equation** of a line in $\mathbb{R}^3$ is defined by a point $P_0(x_0, y_0, z_0)$ and a direction vector $\vec{d} = (a, b, c)$. The parametric equation of the line is given by:

$$\vec{r}(t) = (x_0 + at, y_0 + bt, z_0 + ct), \quad t \in \mathbb{R}$$

This form is useful for describing the position of a point on the line as the parameter t varies.

■ **Example 4.9** For the line passing through the point $P_0(1,2,3)$ with a direction vector $\vec{d} = (4,-1,2)$, the parametric equation is:

$$\vec{r}(t) = (1+4t, 2-t, 3+2t), \quad t \in \mathbb{R}$$

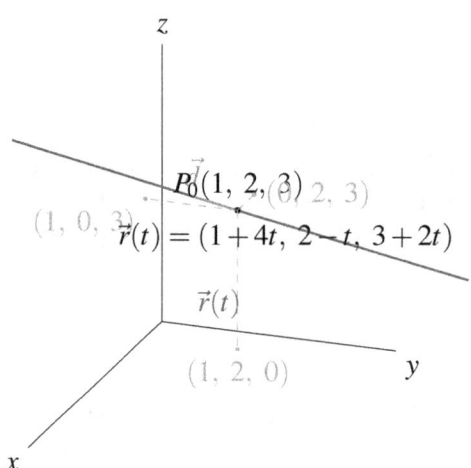

Figura 4.3.1: *Graphical representation of the line passing through $P_0(1,2,3)$ with direction vector $\vec{d} = (4,-1,2)$ in $\mathbb{R}^3$.*

Lema 4.3.1 If two lines in $\mathbb{R}^3$ have parallel direction vectors, the lines are either parallel or coincident. Two vectors $\vec{d_1}$ and $\vec{d_2}$ are parallel if there exists a scalar k such that $\vec{d_1} = k\vec{d_2}$.

Demostración. Let us consider two lines in $\mathbb{R}^3$, each defined by a direction vector, $\vec{d_1}$ for line L_1 and $\vec{d_2}$ for line L_2. If these direction vectors are parallel, then there exists a scalar k such that $\vec{d_1} = k\vec{d_2}$.

4.3 Parametric Equations and Cartesian Equations of Lines and Planes

This implies that both vectors share the same direction, so the lines L_1 and L_2 align in the same direction in three-dimensional space. Hence, the two lines may either be parallel, meaning they never intersect, or coincident if they lie on the same line.
Therefore, if the direction vectors are parallel, the lines will either be parallel or coincident. ∎

> **Theorem 4.3.1** Two lines in $\mathbb{R}^3$ are coincident if they share a point and have parallel direction vectors. Formally, if the lines are defined by:
> $$\vec{r}_1(t) = \vec{P}_0 + t\vec{d}_1, \quad \vec{r}_2(s) = \vec{Q}_0 + s\vec{d}_2$$
> and $\vec{d}_1 = k\vec{d}_2$ for some $k \neq 0$, then the lines are coincident if there exist values t and s such that $\vec{r}_1(t) = \vec{r}_2(s)$.

> **Exercise 4.11** Determine the parametric equation of the line passing through the point $P_0(2,-1,4)$ with direction vector $\vec{d} = (-1,3,5)$.

> **Exercise 4.12** Verify whether the lines $\vec{r}_1(t) = (1+2t, -3+t, 4-t)$ and $\vec{r}_2(s) = (3-s, -2+0{,}5s, 5-0{,}5s)$ are parallel, coincident, or intersect.

4.3.2 Cartesian Equations of a Line

> **Definition 4.3.2** The **Cartesian equation** of a line in $\mathbb{R}^2$ is defined as a linear relationship between the coordinates x and y. The point-slope form of the equation of a line passing through the point (x_1, y_1) with slope m is:
> $$y - y_1 = m(x - x_1)$$
> This equation describes all points that belong to the line with slope m.

■ **Example 4.10** Given the line passing through the point $(2,3)$ with a slope of $m = 4$, the Cartesian equation is:
$$y - 3 = 4(x - 2) \implies y = 4x - 5$$
∎

> **Lema 4.3.2** The slope of a line can be interpreted as the **rate of change** of y with respect to x. If the slope is positive, the line rises from left to right; if it is negative, the line falls from left to right.

> **Theorem 4.3.2** Two lines are **parallel** if and only if they have the same slope. Formally, if the lines are defined by $y = m_1 x + b_1$ and $y = m_2 x + b_2$, then the lines are parallel if $m_1 = m_2$.

Demostración. To prove the theorem, consider two lines defined by the equations:
$$y = m_1 x + b_1 \quad \text{and} \quad y = m_2 x + b_2$$

where m_1 and m_2 are the slopes of each line, and b_1 and b_2 are the y-intercepts.
Two lines are parallel if they have the same inclination and never intersect. This happens if and only if their slopes are equal. In other words, if $m_1 = m_2$, the lines have the same direction and are therefore parallel.

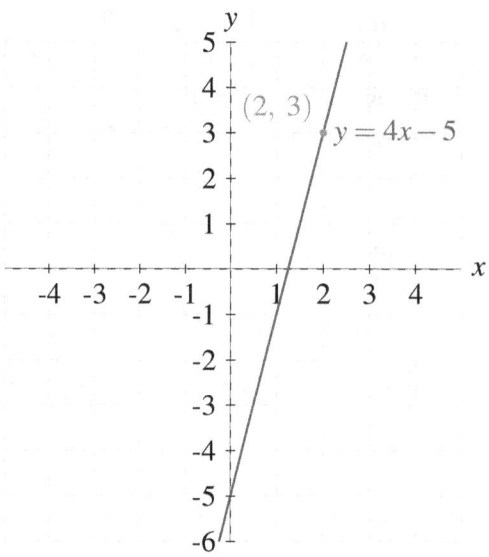

Figura 4.3.2: *Graph of the line $y = 4x - 5$ passing through the point $(2,3)$.*

On the other hand, if $m_1 \neq m_2$, the lines have different inclinations, meaning they will eventually intersect at some point, and thus are not parallel.

In conclusion, the lines are parallel if and only if $m_1 = m_2$. ∎

> **Corollary 4.3.3** Two lines are **perpendicular** if the product of their slopes is -1. If the slopes of the lines are m_1 and m_2, then the lines are perpendicular if:
>
> $$m_1 \cdot m_2 = -1$$

Demostración. To prove the corollary, consider two lines with slopes m_1 and m_2. We want to show that the lines are perpendicular if and only if the product of their slopes equals -1.

Suppose the lines are perpendicular. This means that the angle between the two lines is $90°$. In analytic geometry, the relationship between the slopes of two perpendicular lines is expressed as:

$$m_1 \cdot m_2 = -1$$

This is because, for two lines to form a right angle, their slopes must have a relationship where one is the negative reciprocal of the other, ensuring perpendicularity.

Conversely, if $m_1 \cdot m_2 = -1$, we can conclude that the lines are perpendicular, as the product of the slopes of two perpendicular lines is always equal to -1.

In conclusion, two lines are perpendicular if and only if the product of their slopes is equal to -1. ∎

> ® The Cartesian equation of a line is particularly useful for identifying the slope and the y-intercept. This is fundamental in solving problems in analytic geometry and optimization.

> **Exercise 4.13** Find the Cartesian equation of the line passing through the points $(1,2)$ and $(3,6)$.

4.3 Parametric Equations and Cartesian Equations of Lines and Planes

Exercise 4.14 Determine whether the lines $y = 2x + 3$ and $y = 2x - 5$ are parallel or perpendicular.

4.3.3 Equation of a Plane

Definition 4.3.3 The **Cartesian equation of a plane** in three-dimensional space $\mathbb{R}^3$ is defined as:

$$a(x - x_0) + b(y - y_0) + c(z - z_0) = 0$$

where (x_0, y_0, z_0) is a point on the plane, and $\vec{n} = (a, b, c)$ is a normal vector to the plane. This equation describes all the points that lie on the plane defined by the point P_0 and the vector $\vec{n}$.

■ **Example 4.11** Consider a plane in $\mathbb{R}^3$ passing through the point $(1, -2, 3)$ with a normal vector $\vec{n} = (2, 1, -1)$. The equation of the plane is:

$$2(x - 1) + 1(y + 2) - 1(z - 3) = 0$$

Simplifying, we obtain:

$$2x + y - z - 3 = 0$$

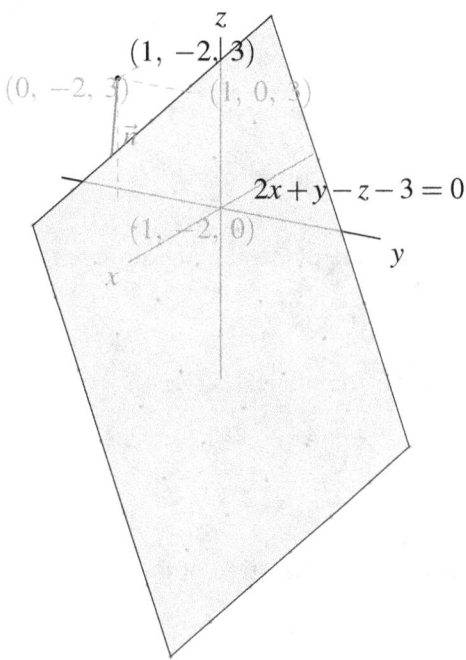

Figura 4.3.3: *Graphical representation of the plane $2x + y - z - 3 = 0$ in $\mathbb{R}^3$, passing through the point $(1, -2, 3)$ with normal vector $\vec{n} = (2, 1, -1)$.*

Lema 4.3.3 If a point $P(x_1, y_1, z_1)$ satisfies the equation $a(x - x_0) + b(y - y_0) + c(z - z_0) = 0$, then the point lies on the plane defined by this equation.

Demostración. To prove the lemma, consider the plane defined by the equation:

$$a(x - x_0) + b(y - y_0) + c(z - z_0) = 0$$

where (x_0, y_0, z_0) is a point on the plane, and (a,b,c) is the normal vector to the plane. Let $P(x_1, y_1, z_1)$ be a point that satisfies the equation.
Substituting the coordinates of the point P into the plane's equation:

$$a(x_1 - x_0) + b(y_1 - y_0) + c(z_1 - z_0) = 0$$

If this equality holds, then the point $P(x_1, y_1, z_1)$ lies on the plane defined by the equation. The equation directly describes all points contained in the plane.
Therefore, if the point $P(x_1, y_1, z_1)$ satisfies the equation of the plane, we conclude that P belongs to the plane. ∎

Theorem 4.3.4 Two planes are **parallel** if their normal vectors are proportional. That is, two planes with normal vectors $\vec{n_1} = (a_1, b_1, c_1)$ and $\vec{n_2} = (a_2, b_2, c_2)$ are parallel if there exists a scalar $\lambda \neq 0$ such that:

$$(a_1, b_1, c_1) = \lambda (a_2, b_2, c_2)$$

Corollary 4.3.5 Two planes are **perpendicular** if the dot product of their normal vectors is zero. Given the normal vectors of two planes $\vec{n_1} = (a_1, b_1, c_1)$ and $\vec{n_2} = (a_2, b_2, c_2)$, the planes are perpendicular if:

$$a_1 a_2 + b_1 b_2 + c_1 c_2 = 0$$

Exercise 4.15 Find the equation of the plane passing through the point $(2, 0, -1)$ and perpendicular to the vector $\vec{n} = (1, -3, 2)$.

Exercise 4.16 Determine whether the planes $2x - y + z = 4$ and $4x - 2y + 2z = 8$ are parallel, perpendicular, or neither.

4.4 Solved Exercises

Exercise 4.17 Find the magnitude and direction of the vector $\vec{v} = (3, 4)$ in $\mathbb{R}^2$.

Demostración. The magnitude of a vector $\vec{v} = (v_1, v_2)$ in $\mathbb{R}^2$ is calculated as:

$$|\vec{v}| = \sqrt{v_1^2 + v_2^2}$$

Substituting the values of $\vec{v} = (3, 4)$:

$$|\vec{v}| = \sqrt{3^2 + 4^2} = \sqrt{9 + 16} = \sqrt{25} = 5$$

To find the direction, we determine the angle θ the vector makes with the x-axis, using:

$$\tan(\theta) = \frac{v_2}{v_1} = \frac{4}{3}$$

Thus:

$$\theta = \tan^{-1}\left(\frac{4}{3}\right)$$

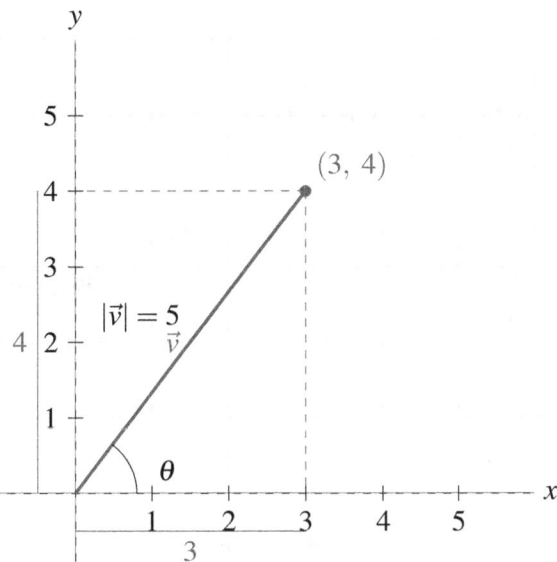

Figura 4.4.1: *Representation of the vector $\vec{v} = (3,4)$ in $\mathbb{R}^2$, showing its magnitude and direction θ.*

4.5 Proposed Exercises

4.5.1 Vector Definitions and Operations

Exercise 4.18 Given the vectors $\vec{u} = (3,-2)$ and $\vec{v} = (-1,5)$, find the sum and difference of the vectors.

Exercise 4.19 Calculate the magnitude of the vector $\vec{w} = (4,-3,1)$ in $\mathbb{R}^3$.

Exercise 4.20 Draw the vectors $\vec{a} = (2,3)$ and $\vec{b} = (-1,-4)$ in $\mathbb{R}^2$, and determine their resulting coordinates when adding the vectors.

Exercise 4.21 Given the vector $\vec{p} = (1,-2,3)$, calculate the product of the vector by a scalar $k = 4$.

Exercise 4.22 Determine if the vectors $\vec{u} = (3,2,-1)$ and $\vec{v} = (-6,-4,2)$ are collinear.

4.5.2 Dot Product and Cross Product

Exercise 4.23 Calculate the dot product of the vectors $\vec{a} = (1,2,3)$ and $\vec{b} = (4,-1,0)$.

Exercise 4.24 Find the angle between the vectors $\vec{u} = (3,4)$ and $\vec{v} = (-4,3)$ in $\mathbb{R}^2$.

Exercise 4.25 Calculate the cross product of the vectors $\vec{p} = (2,-1,3)$ and $\vec{q} = (0,1,4)$.

Exercise 4.26 Determine if the vectors $\vec{a} = (1,0,-1)$ and $\vec{b} = (-1,0,1)$ are orthogonal using the dot product.

Exercise 4.27 Calculate the area of the parallelogram formed by the vectors $\vec{u} = (1,2,0)$ and $\vec{v} = (-3,4,1)$ using the cross product.

4.5.3 Parametric and Cartesian Equations of Lines and Planes

Exercise 4.28 Find the parametric equations of the line passing through the point $A(1,2,3)$ with direction vector $\vec{d} = (4,5,6)$.

Exercise 4.29 Determine the Cartesian equation of the line passing through the points $P_1(2,-1)$ and $P_2(5,3)$ in $\mathbb{R}^2$.

Exercise 4.30 Find the equation of the plane passing through the point $P_0(1,-2,3)$ and perpendicular to the normal vector $\vec{n} = (2,-1,4)$.

Exercise 4.31 Given the point $P(2,1,0)$, determine whether it lies on the line with parametric equations $x = 1+t, y = -2+3t, z = 4-t$.

II Analytic Geometry and Functions

5   Equations of Lines and Planes 105
- 5.1   General Equation and Parametric Equation of a Line
- 5.2   Conditions for Parallelism and Perpendicularity
- 5.3   Equation of a Plane in Space
- 5.4   Solved Exercises
- 5.5   Proposed Exercises

6   Equations of Conics 127
- 6.1   Circle: Equation and Properties
- 6.2   Parabola: Equation and Applications
- 6.3   Ellipse and Hyperbola: Definition and Characteristics
- 6.4   Solved Exercises
- 6.5   Proposed Exercises

7   Functions: Domain, Range, and Operations   145
- 7.1   Domain and Range of Real Functions
- 7.2   Operations with Functions (Addition, Subtraction, Multiplication, Division)
- 7.3   Piecewise-Defined Functions
- 7.4   Solved Exercises
- 7.5   Proposed Exercises

8   Graphs and Graphical Transformations   163
- 8.1   Translation and Reflection of Functions
- 8.2   Vertical and Horizontal Scaling
- 8.3   Graphs of Composite Functions
- 8.4   Solved Exercises
- 8.5   Proposed Exercises

9   Composition and Inverse Functions .. 183
- 9.1   Function Composition and Its Notation
- 9.2   Inverse of a Function and Its Calculation
- 9.3   Properties of Inverse Functions
- 9.4   Solved Exercises
- 9.5   Proposed Exercises

5. Equations of Lines and Planes

5.1 General Equation and Parametric Equation of a Line

5.1.1 General Equation of a Line in the Plane

Definition 5.1.1 The **general equation of a line in the plane** is an equation of the form $Ax + By + C = 0$, where A, B, and C are real constants, and (x, y) represents a point in the Cartesian plane. This equation describes all lines in the plane, as long as at least one of the coefficients A or B is not zero.

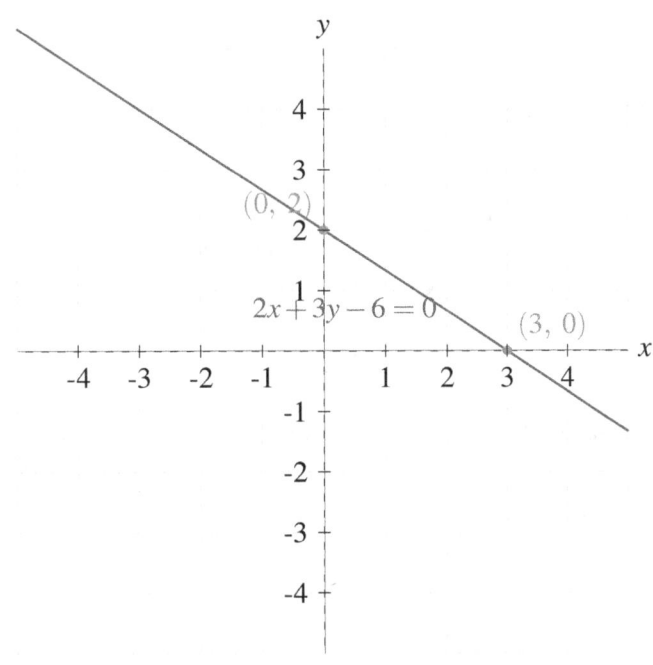

Figura 5.1.1: *Graphical representation of the general equation of a line in the plane* $2x + 3y - 6 = 0$.

Capítulo 5. Equations of Lines and Planes

■ **Example 5.1** Consider the equation $2x - 3y + 6 = 0$. This is a general equation of a line. We can solve for y to obtain its slope-intercept form:

$$y = \frac{2}{3}x + 2$$

This allows us to visualize the slope and the y-intercept of the line. In this case, the slope is $\frac{2}{3}$, and the y-intercept is 2. ■

Lema 5.1.1 For the equation $Ax + By + C = 0$ with $A, B, C \in \mathbb{R}$ and $A^2 + B^2 \neq 0$, the slope of the line can be expressed as $m = -\frac{A}{B}$ provided $B \neq 0$.

Demostración. We solve for y in the general equation to obtain the slope-intercept form:

$$Ax + By + C = 0 \quad \Rightarrow \quad y = -\frac{A}{B}x - \frac{C}{B}, \quad \text{if } B \neq 0$$

Thus, the slope m of the line is $m = -\frac{A}{B}$. ∎

Theorem 5.1.1 Two lines in the plane given by the equations $A_1x + B_1y + C_1 = 0$ and $A_2x + B_2y + C_2 = 0$ are **parallel** if and only if $\frac{A_1}{B_1} = \frac{A_2}{B_2}$.

Demostración. For two lines to be parallel, they must have the same slope. The slope of the first line is $m_1 = -\frac{A_1}{B_1}$, and the slope of the second line is $m_2 = -\frac{A_2}{B_2}$. The slopes will be equal if:

$$-\frac{A_1}{B_1} = -\frac{A_2}{B_2} \quad \Rightarrow \quad \frac{A_1}{B_1} = \frac{A_2}{B_2}$$

This proves the condition for the lines to be parallel. ∎

Corollary 5.1.2 If two lines in the plane are perpendicular, then the product of their slopes is -1. That is, if the equations of the lines are $A_1x + B_1y + C_1 = 0$ and $A_2x + B_2y + C_2 = 0$, then:

$$\left(-\frac{A_1}{B_1}\right) \cdot \left(-\frac{A_2}{B_2}\right) = -1$$

Demostración. To prove that two lines are perpendicular if the product of their slopes is -1, we first find the slopes of both lines from their general equations.
The slope of a line given by $A_1x + B_1y + C_1 = 0$ is:

$$m_1 = -\frac{A_1}{B_1}$$

Similarly, the slope of the second line, whose equation is $A_2x + B_2y + C_2 = 0$, is:

$$m_2 = -\frac{A_2}{B_2}$$

The lines will be perpendicular if the product of their slopes equals -1:

$$m_1 \cdot m_2 = \left(-\frac{A_1}{B_1}\right)\left(-\frac{A_2}{B_2}\right) = \frac{A_1 A_2}{B_1 B_2} = -1$$

Thus, if the product of the slopes is -1, the lines are perpendicular. ∎

5.1 General Equation and Parametric Equation of a Line

■ **Example 5.2** Consider the lines given by the equations $3x+4y-12=0$ and $4x-3y+5=0$.
To verify if these lines are perpendicular, we calculate their slopes:
First, rewrite the equations in slope-intercept form ($y = mx+b$).
For the first line:

$$3x+4y-12=0$$
$$4y = -3x+12$$
$$y = -\frac{3}{4}x+3$$

Thus, the slope is $m_1 = -\frac{3}{4}$.
For the second line:

$$4x-3y+5=0$$
$$-3y = -4x-5$$
$$y = \frac{4}{3}x+\frac{5}{3}$$

Thus, the slope is $m_2 = \frac{4}{3}$.
The product of the slopes is:

$$m_1 \cdot m_2 = \left(-\frac{3}{4}\right)\left(\frac{4}{3}\right) = -1$$

Therefore, the lines are perpendicular. ■

> **R** The general equation of a line provides a versatile representation of lines in the plane, useful for analyzing relationships such as parallelism and perpendicularity. Moreover, this equation is particularly convenient for working with line intersections or general analytic geometry problems.

Exercise 5.1 Given the equation of the line $5x-2y+10=0$, find its slope and y-intercept. ■

Exercise 5.2 Determine whether the lines given by the equations $x-y+1=0$ and $2x-2y-3=0$ are parallel, perpendicular, or neither. ■

Exercise 5.3 Find the equation of the line passing through the point $(1,-2)$ and parallel to the line $3x-4y+7=0$. ■

Exercise 5.4 Calculate the intersection point of the lines $x+y-5=0$ and $2x-y+3=0$. ■

Exercise 5.5 Determine the distance from the point $(3,4)$ to the line $6x-8y+5=0$. Use the formula for the distance from a point to a line. ■

> **R** The proposed exercises allow for practical application of the concepts studied, from finding slopes to determining intersections and parallelism. Each new result connects with the previous one, reinforcing the overall understanding of line behavior in the plane.

5.1.2 Parametric and Vector Equations

Definition 5.1.2 A **parametric equation** of a line in space or the plane is expressed using a parameter, typically denoted as t, and describes each point on the line as a function of that parameter. In the plane, the parametric equation of a line can be written as:

$$x = x_0 + ta_1, \quad y = y_0 + ta_2$$

where (x_0, y_0) is a point on the line, and (a_1, a_2) is a direction vector of the line.

Definition 5.1.3 The **vector equation** of a line is a representation in the form of a vector describing all points on the line. In the plane, the vector equation can be written as:

$$\vec{r} = \vec{r_0} + t\vec{v}$$

where $\vec{r_0}$ is the position vector of a point on the line, and $\vec{v}$ is the direction vector.

■ **Example 5.3** Consider the vector equation of a line $\vec{r} = (2, 3) + t(1, -2)$. If we take $t = 3$, we find a point on the line:

$$\vec{r} = (2, 3) + 3(1, -2) = (2 + 3, 3 - 6) = (5, -3)$$

■

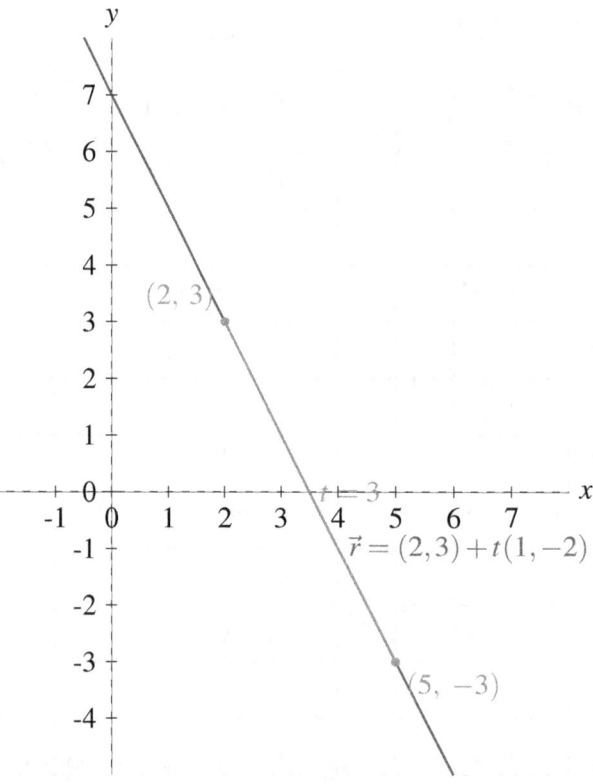

Figura 5.1.2: *Graph of the line $\vec{r} = (2, 3) + t(1, -2)$, showing the points $(2, 3)$ and $(5, -3)$ for $t = 0$ and $t = 3$.*

Lema 5.1.2 If a line is defined by the parametric equations $x = x_0 + ta_1$, $y = y_0 + ta_2$, any point (x_1, y_1) on the line satisfies these equations for some value of t.

5.1 General Equation and Parametric Equation of a Line

Demostración. Given a point (x_1, y_1) on the line, there exists a value of t such that:

$$x_1 = x_0 + ta_1, \quad y_1 = y_0 + ta_2$$

Solving for t, we demonstrate that (x_1, y_1) belongs to the line. ∎

Theorem 5.1.3 The parametric and vector equations of a line are equivalent in the sense that both describe the same set of points.

Demostración. Consider the vector equation $\vec{r} = \vec{r_0} + t\vec{v}$. Decomposing $\vec{r}$, $\vec{r_0}$, and $\vec{v}$ into their components, we obtain:

$$(x, y) = (x_0, y_0) + t(a_1, a_2)$$

which translates to the parametric equations:

$$x = x_0 + ta_1, \quad y = y_0 + ta_2$$

Therefore, both equations describe the same set of points, proving their equivalence. ∎

Corollary 5.1.4 If a line has a vector equation $\vec{r} = \vec{r_0} + t\vec{v}$, then any point on the line can be described as a linear combination of the vectors $\vec{r_0}$ and $\vec{v}$.

Demostración. Given the vector equation of the line:

$$\vec{r} = \vec{r_0} + t\vec{v}$$

where $\vec{r_0}$ is a fixed point on the line, $\vec{v}$ is a direction vector, and t is a real parameter.
We can write any point on the line $\vec{r}$ as:

$$\vec{r} = \vec{r_0} + t\vec{v} = 1 \cdot \vec{r_0} + t \cdot \vec{v}$$

This implies that the point $\vec{r}$ can be expressed as a linear combination of the vectors $\vec{r_0}$ and $\vec{v}$, with coefficients 1 and t, respectively.
Thus, any point on the line is described as a linear combination of the vectors $\vec{r_0}$ and $\vec{v}$. ∎

Exercise 5.6 Find the parametric equation of the line passing through the points $(1, 4)$ and $(3, 0)$. Represent the equation also in vector form. ∎

Exercise 5.7 Determine if the point $(5, -1)$ lies on the line with vector equation $\vec{r} = (2, 3) + t(1, -2)$. ∎

Exercise 5.8 Calculate the value of t for which the point on the line with parametric equation $x = 2 + t$, $y = -1 + 2t$ has coordinates $(4, 3)$. ∎

Exercise 5.9 Find the parametric equation of the line perpendicular to the line $x = 1 + 2t$, $y = 3 - t$ and passing through the point $(0, 0)$. ∎

> (R) The provided exercises aim to reinforce the connection between the different representations of a line, whether in parametric or vector form. They also allow analysis of fundamental concepts such as perpendicularity and the inclusion of a point on a line.

5.1.3 Conversion Between Forms of the Equation

Definition 5.1.4 The **conversion between forms of a line equation** refers to the process of changing the representation of a line from one specific form (such as general form, point-slope form, or parametric form) to another. This process allows for analyzing and solving analytic geometry problems using the most convenient form for a given situation.

■ **Example 5.4** Consider the equation of a line in point-slope form: $y - 2 = 3(x - 1)$. We can convert this equation to general form:

$$y - 2 = 3x - 3$$

Rearranging the terms to obtain the general form:

$$3x - y - 1 = 0$$

In this example, we see how the equation changes from a form that is convenient for identifying the slope and a point on the line to the general form, which is useful for algebraic analysis. ■

Lema 5.1.3 The general form of a line equation, $Ax + By + C = 0$, can be converted to slope-intercept form $y = mx + b$ if $B \neq 0$, where $m = -\frac{A}{B}$ and $b = -\frac{C}{B}$.

Demostración. To convert $Ax + By + C = 0$ to slope-intercept form, solve for y:

$$By = -Ax - C$$

$$y = -\frac{A}{B}x - \frac{C}{B}$$

Thus, the slope m is $-\frac{A}{B}$, and the y-intercept b is $-\frac{C}{B}$. ∎

Theorem 5.1.5 The parametric and vector forms of a line in the plane can be converted into each other equivalently and consistently. The conversion between these forms does not change the set of points represented by the line.

Demostración. Consider a line defined by the parametric equations:

$$x = x_0 + ta_1, \quad y = y_0 + ta_2$$

We can write the same line in vector form:

$$\vec{r} = \vec{r_0} + t\vec{v}$$

where $\vec{r_0} = (x_0, y_0)$ and $\vec{v} = (a_1, a_2)$. The expressions for x and y in the parametric equation are simply the components of the vector $\vec{r}$, which proves the equivalence between the forms. ∎

5.1 General Equation and Parametric Equation of a Line

Corollary 5.1.6 If a line in the plane is expressed in general form $Ax + By + C = 0$ and we wish to convert it to parametric form, we first solve for one of the variables (e.g., x or y) and then define a parameter t to traverse the points on the line.

■ **Example 5.5** Let us convert the general equation $2x + 3y - 6 = 0$ to its parametric form. First, solve for y:

$$3y = -2x + 6$$

$$y = -\frac{2}{3}x + 2$$

Now, define a parameter $t = x$:

$$x = t, \quad y = -\frac{2}{3}t + 2$$

Thus, the parametric form of the line is:

$$(x, y) = (t, -\frac{2}{3}t + 2), \quad t \in \mathbb{R}$$

■

⊙ Converting between different forms of a line equation is essential for applying appropriate resolution techniques in various analytic geometry problems. For instance, the slope-intercept form is convenient when interpreting the slope and y-intercept, while the parametric form facilitates the interpretation of specific points along the line.

Exercise 5.10 Given the equation $x - 2y + 4 = 0$, convert it to slope-intercept form and identify the slope and y-intercept.

Exercise 5.11 Convert the general equation $3x + 4y - 12 = 0$ to its parametric form and graph the line.

Exercise 5.12 Find the vector form of the line passing through the point $(2, -1)$ with direction vector $\vec{v} = (1, 3)$.

Exercise 5.13 Given the parametric form $x = 1 + t$, $y = 2 - t$, convert this equation to the general form of the line.

Exercise 5.14 Verify if the point $(4, -2)$ lies on the line with the vector equation $\vec{r} = (1, 2) + t(3, -4)$.

⊙ The provided exercises allow the reader to practice converting between different forms of a line equation, ensuring a solid understanding of analytic geometry. This skill is crucial when solving problems that require the most appropriate representation of a line.

5.2 Conditions for Parallelism and Perpendicularity

5.2.1 Parallel and Perpendicular Vectors

Definition 5.2.1 Two **vectors are parallel** if one is a scalar multiple of the other. That is, given vectors $\vec{a}$ and $\vec{b}$, they are parallel if there exists a real number λ such that $\vec{a} = \lambda \vec{b}$. **Perpendicular vectors**, on the other hand, are those whose dot product is zero, i.e., $\vec{a} \cdot \vec{b} = 0$.

■ **Example 5.6** Consider the vectors $\vec{a} = (2,4)$ and $\vec{b} = (1,2)$. We can see that $\vec{a} = 2\vec{b}$, which indicates that $\vec{a}$ and $\vec{b}$ are parallel. Now, if we consider the vectors $\vec{c} = (3,4)$ and $\vec{d} = (-4,3)$, we calculate their dot product:

$$\vec{c} \cdot \vec{d} = 3(-4) + 4(3) = -12 + 12 = 0$$

Therefore, $\vec{c}$ and $\vec{d}$ are perpendicular. ■

Lema 5.2.1 If the vectors $\vec{a} = (a_1, a_2)$ and $\vec{b} = (b_1, b_2)$ are parallel, then $a_1 b_2 = a_2 b_1$. This condition is sufficient to verify if two vectors in $\mathbb{R}^2$ are parallel.

Demostración. If $\vec{a}$ and $\vec{b}$ are parallel, then $\vec{a} = \lambda \vec{b}$ for some scalar λ. This implies:

$$(a_1, a_2) = \lambda(b_1, b_2) \quad \Rightarrow \quad a_1 = \lambda b_1, \quad a_2 = \lambda b_2$$

Dividing the two equations gives:

$$\frac{a_1}{b_1} = \frac{a_2}{b_2} = \lambda$$

Therefore, $a_1 b_2 = a_2 b_1$. ■

Theorem 5.2.1 Two vectors $\vec{a}$ and $\vec{b}$ are perpendicular if and only if their dot product is zero. That is, $\vec{a} \cdot \vec{b} = 0$ implies that $\vec{a}$ and $\vec{b}$ are perpendicular.

Demostración. The dot product of two vectors $\vec{a} = (a_1, a_2)$ and $\vec{b} = (b_1, b_2)$ is:

$$\vec{a} \cdot \vec{b} = a_1 b_1 + a_2 b_2$$

If $\vec{a}$ and $\vec{b}$ are perpendicular, then:

$$a_1 b_1 + a_2 b_2 = 0$$

This implies that there is no projection of one vector onto the other, which means they form a 90° angle and are therefore perpendicular. ■

Corollary 5.2.2 In $\mathbb{R}^3$, two vectors $\vec{a} = (a_1, a_2, a_3)$ and $\vec{b} = (b_1, b_2, b_3)$ are perpendicular if the following condition holds:

$$a_1 b_1 + a_2 b_2 + a_3 b_3 = 0$$

5.2 Conditions for Parallelism and Perpendicularity

This extends the condition for perpendicularity from two dimensions to three dimensions.

■ **Example 5.7** Consider the vectors $\vec{a} = (1, 2, -1)$ and $\vec{b} = (2, -1, 1)$ in $\mathbb{R}^3$. We calculate their dot product:

$$\vec{a} \cdot \vec{b} = 1 \cdot 2 + 2 \cdot (-1) + (-1) \cdot 1 = 2 - 2 - 1 = -1$$

Since the dot product is not zero, $\vec{a}$ and $\vec{b}$ are not perpendicular. ■

> (R) The conditions for parallelism and perpendicularity between two vectors are fundamental in geometry and linear algebra, especially when understanding the angular relationship between different directions in space.

Exercise 5.15 Determine whether the vectors $\vec{u} = (3, -6)$ and $\vec{v} = (-1, 2)$ are parallel, perpendicular, or neither. Graphically represent both vectors. ■

Exercise 5.16 Find a vector $\vec{w}$ in $\mathbb{R}^2$ that is perpendicular to the vector $\vec{a} = (4, 1)$. Graphically represent the vectors $\vec{a}$ and $\vec{w}$. ■

> (R) The proposed exercises are essential for solidifying the concept of parallelism and perpendicularity, enabling the reader to identify the relationship between different vectors both algebraically and geometrically.

5.2.2 Calculating the Slope to Determine Parallelism

Definition 5.2.2 The **slope** of a line is a value that describes the inclination of the line relative to the x-axis. Given a line passing through the points (x_1, y_1) and (x_2, y_2), the slope m is calculated using the formula:

$$m = \frac{y_2 - y_1}{x_2 - x_1}$$

Two lines are **parallel** if and only if their slopes are equal, i.e., $m_1 = m_2$.

■ **Example 5.8** Consider the lines L_1 and L_2 passing through the points $(1, 2)$ and $(3, 6)$, and $(0, 1)$ and $(2, 5)$, respectively. Let's calculate the slope of each line:
For L_1:

$$m_1 = \frac{6-2}{3-1} = \frac{4}{2} = 2$$

For L_2:

$$m_2 = \frac{5-1}{2-0} = \frac{4}{2} = 2$$

Since $m_1 = m_2 = 2$, we conclude that the lines L_1 and L_2 are parallel. ■

Lema 5.2.2 If two lines have equal slopes, then the lines are parallel, provided they do not coincide. This implies that equal slopes ensure the lines never intersect unless they are the same line.

Capítulo 5. Equations of Lines and Planes

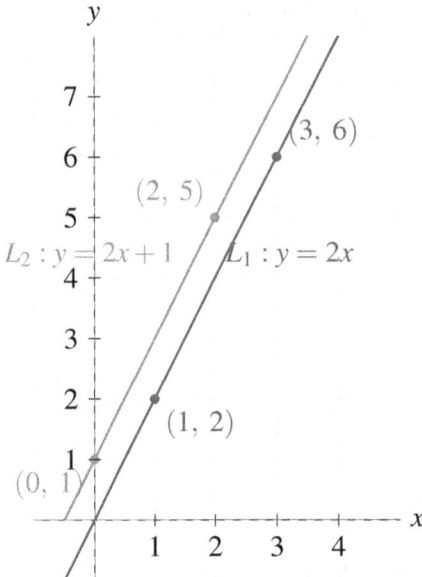

Figura 5.2.1: *Graph of lines L_1 and L_2, showing they are parallel with $m_1 = m_2 = 2$.*

Demostración. Let L_1 and L_2 be two lines with slopes m_1 and m_2. If $m_1 = m_2$, it means both lines have the same inclination with respect to the x-axis. Thus, they are either parallel (not intersecting) or coincident (sharing all their points). ∎

> **Theorem 5.2.3** The lines $Ax + By + C = 0$ and $A'x + B'y + C' = 0$ are parallel if and only if $\frac{A}{B} = \frac{A'}{B'}$, provided $B, B' \neq 0$.

Demostración. Since the slope m of a line in general form $Ax + By + C = 0$ is $m = -\frac{A}{B}$, for the lines L_1 and L_2 to be parallel, their slopes must be equal:

$$-\frac{A}{B} = -\frac{A'}{B'}$$

Eliminating the negative sign, we have:

$$\frac{A}{B} = \frac{A'}{B'}$$

Thus, the lines are parallel if this equality holds. ∎

> **Corollary 5.2.4** If two lines are parallel, then the distance between them is constant throughout the plane. This means that for any point on one line, the shortest distance to the other line is always the same.

■ **Example 5.9** Consider the lines $L_1 : 3x - 4y + 5 = 0$ and $L_2 : 6x - 8y - 10 = 0$. To verify if they are parallel, we calculate the slope of each:
For L_1:

$$m_1 = -\frac{A}{B} = -\frac{3}{-4} = \frac{3}{4}$$

5.2 Conditions for Parallelism and Perpendicularity

For L_2:

$$m_2 = -\frac{A'}{B'} = -\frac{6}{-8} = \frac{3}{4}$$

Since $m_1 = m_2$, we conclude that the lines L_1 and L_2 are parallel. ∎

> (R) Calculating the slope is a fundamental tool for determining whether two lines are parallel. This property is particularly useful in solving analytic geometry problems that require verifying the relative orientation of different lines in the plane.

> **Exercise 5.17** Given the line $y = 2x + 3$, find the equation of another line parallel to it that passes through the point $(1, -2)$.

> **Exercise 5.18** Verify whether the lines passing through the points $(0,1)$ and $(2,5)$, and $(1,2)$ and $(3,6)$, respectively, are parallel.

> (R) These exercises allow the reader to apply the concepts of slope calculation and parallelism, reinforcing practical understanding of how to determine whether two lines are parallel based on their equations or the points that define them.

5.2.3 Applications in Solving Geometric Problems

> **Definition 5.2.3** The **solution of geometric problems** involves using principles and properties of geometry, such as slopes, line equations, and angular relationships, to determine specific characteristics of geometric figures or points in the plane. These applications include determining whether two lines are parallel or perpendicular, calculating distances and areas, and finding intersection points.

■ **Example 5.10** Consider a triangle defined by the points $A = (1,2)$, $B = (4,6)$, and $C = (7,2)$. To verify if the triangle is isosceles, we calculate the distances between each pair of points:

$$AB = \sqrt{(4-1)^2 + (6-2)^2} = \sqrt{3^2 + 4^2} = \sqrt{9+16} = 5$$

$$BC = \sqrt{(7-4)^2 + (2-6)^2} = \sqrt{3^2 + (-4)^2} = \sqrt{9+16} = 5$$

$$AC = \sqrt{(7-1)^2 + (2-2)^2} = \sqrt{6^2} = 6$$

Since $AB = BC = 5$, we conclude that the triangle ABC is isosceles.

∎

Lema 5.2.3 If two lines are perpendicular, then the triangle formed by joining them with a third line parallel to one of them is a right triangle.

Demostración. Let L_1 and L_2 be two perpendicular lines with $m_1 \cdot m_2 = -1$. Adding a third line L_3 parallel to L_1, such that $m_3 = m_1$, creates a right triangle since L_1 and L_2 form a 90° angle, and L_3 is parallel to L_1. ∎

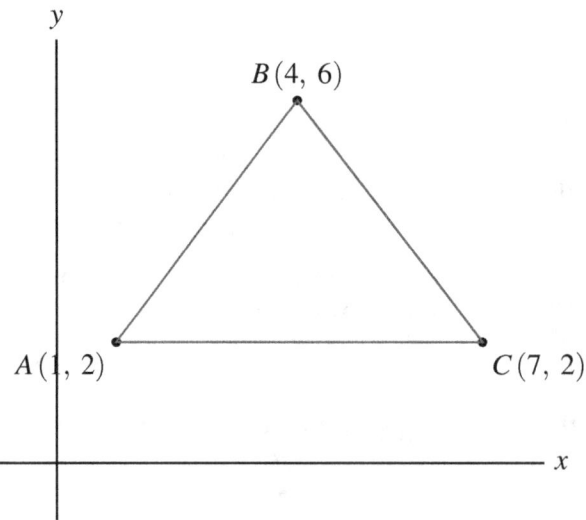

Figura 5.2.2: *Graph of the isosceles triangle ABC in the coordinate plane*

Theorem 5.2.5 In any triangle, the median from a vertex to the midpoint of the opposite side divides the triangle into two triangles of equal area.

Demostración. Consider a triangle ABC with the median AD reaching the midpoint D of side BC. The median divides ABC into two triangles ABD and ADC. Since D is the midpoint of BC, the segments BD and DC are equal. Additionally, both triangles share the same height from vertex A to base BC, implying that their areas are equal. ∎

Corollary 5.2.6 In an equilateral triangle, any median is also an altitude and a bisector. This means that each median divides the triangle into two congruent triangles.

■ **Example 5.11** Suppose an equilateral triangle with sides of length 6. The median from a vertex to the midpoint of the opposite side, also serving as an altitude, divides the triangle into two right triangles with legs 3 and $\sqrt{6^2 - 3^2} = \sqrt{27} = 3\sqrt{3}$. ∎

> (R) The medians, altitudes, and bisectors of a triangle play a fundamental role in solving geometric problems as they allow convenient division of figures, simplifying calculations of areas, perimeters, and other properties.

Exercise 5.19 Given a triangle with vertices at $A = (0,0)$, $B = (6,0)$, and $C = (3,4)$, find the length of the median from vertex C to the midpoint of side AB.

Exercise 5.20 Verify whether the point $(4,3)$ lies on the line passing through points $(1,1)$ and $(5,5)$ and determine if the point divides the segment into two equal parts.

> (R) The proposed exercises allow the reader to reinforce the use of medians, altitudes, and bisectors in geometric problems, providing a deeper understanding of how to apply the properties of geometric figures to solve complex problems.

5.3 Equation of a Plane in Space

5.3.1 Definition and Forms of the Plane Equation

Definition 5.3.1 A **plane** in three-dimensional space $\mathbb{R}^3$ is defined as the set of all points (x, y, z) that satisfy a linear equation of the form:

$$Ax + By + Cz + D = 0$$

where A, B, C, and D are constants, and at least one of A, B, or C is nonzero. This equation is known as the **general form of the plane equation**.

■ **Example 5.12** Consider the plane defined by the equation $2x - y + 3z - 6 = 0$. To verify if the point $P = (1, 0, 2)$ lies on the plane, substitute the coordinates of P into the plane equation:

$$2(1) - 0 + 3(2) - 6 = 2 + 6 - 6 = 2$$

Since the result is not equal to zero, we conclude that the point P does not lie on the plane.

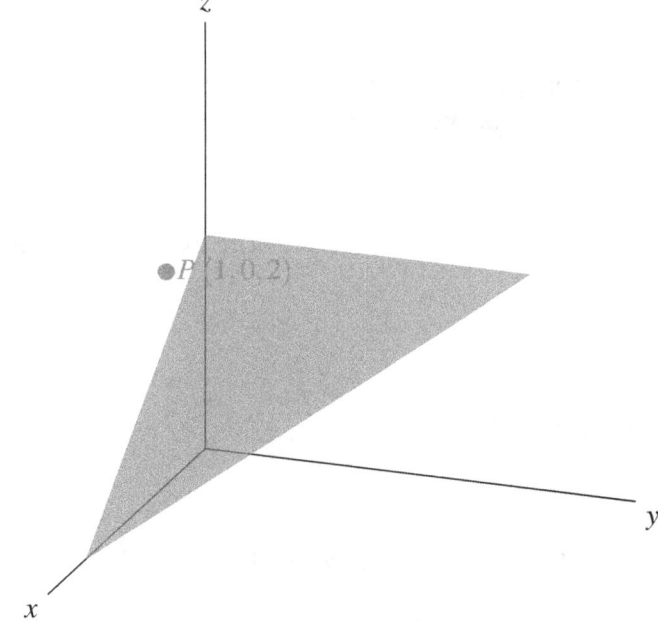

Figura 5.3.1: *Graphical representation of the plane $2x - y + 3z - 6 = 0$ and the point P in three-dimensional space*

■

Lema 5.3.1 If a plane has an equation of the form $Ax + By + Cz + D = 0$, then the vector $\vec{n} = (A, B, C)$ is a **normal vector** to the plane. This vector is perpendicular to any vector contained in the plane.

Demostración. Let $\vec{r} = (x, y, z)$ be a point on the plane, and $\vec{r_0} = (x_0, y_0, z_0)$ be another point also on the plane. Any vector $\vec{v}$ contained in the plane can be expressed as:

$$\vec{v} = \vec{r} - \vec{r_0} = (x - x_0, y - y_0, z - z_0)$$

Since the plane is defined by $Ax+By+Cz+D=0$, the normal vector $\vec{n}=(A,B,C)$ is perpendicular to any vector in the plane, which is true if:

$$\vec{n}\cdot\vec{v}=A(x-x_0)+B(y-y_0)+C(z-z_0)=0$$

Thus, $\vec{n}$ is a normal vector to the plane. ∎

> **Theorem 5.3.1** Any plane in three-dimensional space can be defined using a point $P_0=(x_0,y_0,z_0)$ on the plane and a normal vector $\vec{n}=(A,B,C)$. The equation of the plane is:
>
> $$A(x-x_0)+B(y-y_0)+C(z-z_0)=0$$
>
> This is known as the **point-normal form of the plane equation**.

Demostración. To derive the point-normal form of a plane equation, consider a given point $P_0=(x_0,y_0,z_0)$ on the plane and a normal vector $\vec{n}=(A,B,C)$, which is perpendicular to the plane. For any other point $P=(x,y,z)$ on the plane, the vector connecting P_0 and P is:

$$\vec{r}=(x-x_0,y-y_0,z-z_0)$$

Since the normal vector $\vec{n}$ is perpendicular to the plane, it must also be perpendicular to the vector $\vec{r}$. This implies that the dot product between $\vec{n}$ and $\vec{r}$ is zero:

$$\vec{n}\cdot\vec{r}=A(x-x_0)+B(y-y_0)+C(z-z_0)=0$$

This is the equation of the plane in its point-normal form, as it uses a point on the plane and a normal vector to define the plane. ∎

> **Corollary 5.3.2** If a plane has a normal vector $\vec{n}=(A,B,C)$ and passes through the origin $(0,0,0)$, then the plane equation simplifies to:
>
> $$Ax+By+Cz=0$$
>
> This is a simplified form of the plane equation when $D=0$.

Demostración. If a plane passes through the origin $(0,0,0)$, we can use the point-normal form of the plane equation. Recall that the point-normal form of a plane is:

$$A(x-x_0)+B(y-y_0)+C(z-z_0)=0$$

Since the plane passes through the origin, the values of x_0, y_0, and z_0 are all zero. Substituting these values, we obtain:

$$A(x-0)+B(y-0)+C(z-0)=0$$

Simplifying:

$$Ax+By+Cz=0$$

Thus, the equation of the plane passing through the origin is simply $Ax+By+Cz=0$. ∎

5.3 Equation of a Plane in Space

Example 5.13 Consider the plane passing through the point $P_0 = (1, -2, 3)$ with a normal vector $\vec{n} = (2, 1, -1)$. The point-normal form of the plane equation is:

$$2(x-1) + 1(y+2) - 1(z-3) = 0$$

Simplifying:

$$2x + y - z - 2 + 2 + 3 = 0$$

$$2x + y - z + 3 = 0$$

> **R** The use of different forms of the plane equation, such as the general or point-normal forms, is essential for solving geometric problems in three-dimensional space. Each form has its advantages depending on the available information, such as specific points or normal vectors.

Exercise 5.21 Given the point $P = (2, -1, 4)$ and the normal vector $\vec{n} = (3, 2, -1)$, find the equation of the plane that passes through P and has $\vec{n}$ as its normal vector.

Exercise 5.22 Verify whether the point $Q = (1, 1, 1)$ lies on the plane defined by the equation $x - 2y + 3z - 4 = 0$.

> **R** The proposed exercises allow the reader to apply the concepts of planes and normal vectors, reinforcing the ability to find and manipulate different forms of the plane equation to solve complex geometric problems.

5.3.2 Parallel and Perpendicular Planes

Definition 5.3.2 Two **planes are parallel** if their normal vectors are parallel, meaning that the normal vectors are proportional. If $\vec{n_1} = (A_1, B_1, C_1)$ and $\vec{n_2} = (A_2, B_2, C_2)$ are the normal vectors of two planes, the planes are parallel if there exists a scalar λ such that $\vec{n_1} = \lambda \vec{n_2}$.
On the other hand, two **planes are perpendicular** if the dot product of their normal vectors is zero, i.e., $\vec{n_1} \cdot \vec{n_2} = 0$.

Example 5.14 Consider the planes $P_1 : 2x - 3y + z = 4$ and $P_2 : 4x - 6y + 2z = 7$. The normal vectors of these planes are $\vec{n_1} = (2, -3, 1)$ and $\vec{n_2} = (4, -6, 2)$. We observe that:

$$\vec{n_2} = 2 \cdot \vec{n_1}$$

Therefore, the planes P_1 and P_2 are parallel.

Lema 5.3.2 If two planes have normal vectors $\vec{n_1}$ and $\vec{n_2}$ such that $\vec{n_1} \cdot \vec{n_2} = 0$, then the planes are perpendicular.

Demostración. Let the normal vectors be $\vec{n_1} = (A_1, B_1, C_1)$ and $\vec{n_2} = (A_2, B_2, C_2)$. If the dot product $\vec{n_1} \cdot \vec{n_2} = A_1 A_2 + B_1 B_2 + C_1 C_2 = 0$, this implies that the normal vectors are orthogonal, which means the planes are perpendicular to each other.

Capítulo 5. Equations of Lines and Planes

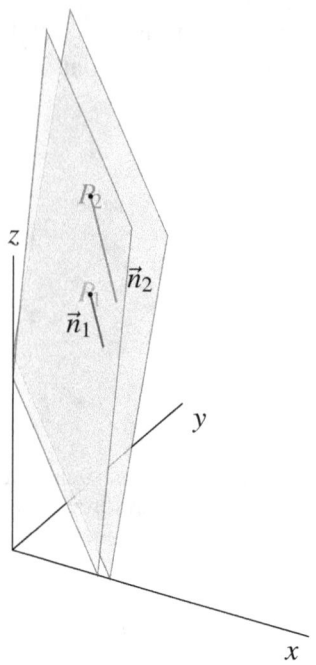

Figura 5.3.2: *Graphical representation of planes $P_1 : 2x - 3y + z = 4$ and $P_2 : 4x - 6y + 2z = 7$, showing that they are parallel as $\vec{n}_2 = 2\vec{n}_1$.*

> **Theorem 5.3.3** Two planes in $\mathbb{R}^3$ are parallel if and only if their equations can be expressed as $A_1x + B_1y + C_1z + D_1 = 0$ and $A_2x + B_2y + C_2z + D_2 = 0$, where the normal vector of the second plane is a scalar multiple of the normal vector of the first plane, i.e., $(A_2, B_2, C_2) = \lambda(A_1, B_1, C_1)$ for some $\lambda \neq 0$.

Demostración. To prove the theorem, consider two planes with equations:

$$a_1x + b_1y + c_1z + d_1 = 0$$

$$a_2x + b_2y + c_2z + d_2 = 0$$

The normal vectors to these planes are, respectively:

$$\vec{n}_1 = (a_1, b_1, c_1), \quad \vec{n}_2 = (a_2, b_2, c_2)$$

Two planes are parallel if their normal vectors are proportional, meaning there exists a scalar $\lambda \neq 0$ such that:

$$\vec{n}_1 = \lambda \vec{n}_2$$

This implies:

$$(a_1, b_1, c_1) = \lambda(a_2, b_2, c_2)$$

This means each component of the normal vector $\vec{n}_1$ is equal to λ times the corresponding component of $\vec{n}_2$:

$$a_1 = \lambda a_2, \quad b_1 = \lambda b_2, \quad c_1 = \lambda c_2$$

When this relationship holds, the planes have proportional normal vectors and, therefore, are parallel to each other. ∎

5.3 Equation of a Plane in Space

Corollary 5.3.4 If two planes are perpendicular and one of them has the equation $Ax + By + Cz + D = 0$, the other plane must have a normal vector satisfying $A'A + B'B + C'C = 0$, where (A', B', C') are the coefficients of the normal vector of the second plane.

Demostración. For two planes to be perpendicular, the dot product of their normal vectors must equal zero. Suppose the first plane has a normal vector $\vec{n}_1 = (A, B, C)$, and the second plane has a normal vector $\vec{n}_2 = (A', B', C')$. The planes are perpendicular if:

$$\vec{n}_1 \cdot \vec{n}_2 = A'A + B'B + C'C = 0$$

A zero dot product indicates that the normal vectors are orthogonal, which implies the planes are perpendicular to each other.

Thus, if one plane has the equation $Ax + By + Cz + D = 0$, the other plane must have a normal vector (A', B', C') such that $A'A + B'B + C'C = 0$ to ensure the perpendicularity of the planes. ∎

■ **Example 5.15** Consider the planes $P_3 : x + 2y - z = 3$ and $P_4 : 2x - y + 2z = 5$. The normal vectors are $\vec{n}_3 = (1, 2, -1)$ and $\vec{n}_4 = (2, -1, 2)$. Calculate their dot product:

$$\vec{n}_3 \cdot \vec{n}_4 = 1 \cdot 2 + 2 \cdot (-1) + (-1) \cdot 2 = 2 - 2 - 2 = -2$$

Since the dot product is not zero, the planes P_3 and P_4 are not perpendicular. ■

(R) The conditions for parallelism and perpendicularity between planes are frequently used in spatial geometry problems to determine the relative orientation of different surfaces. These concepts simplify complex problems by reducing them to analyses of normal vectors.

Exercise 5.23 Find the equation of a plane parallel to the plane $3x - y + 2z - 5 = 0$ that passes through the point $(1, 2, 3)$.

Exercise 5.24 Verify if the planes $x + y + z = 4$ and $2x - y + 2z = 7$ are perpendicular.

(R) The exercises above allow the reader to practice determining parallelism and perpendicularity between planes, using both their equations and normal vectors, which is essential for solving geometric problems in three-dimensional space.

5.3.3 Intersection of Planes

Definition 5.3.3 The **intersection of two planes** in three-dimensional space is, in general, a line. Given two planes $P_1 : A_1 x + B_1 y + C_1 z + D_1 = 0$ and $P_2 : A_2 x + B_2 y + C_2 z + D_2 = 0$, their intersection can be represented as a set of points that satisfy both equations simultaneously. If the planes are not parallel, they intersect in a line.

■ **Example 5.16** Consider the planes $P_1 : x + y + z = 3$ and $P_2 : 2x - y + z = 1$. To find the equation of the line of intersection, we solve the system of equations simultaneously. Subtracting the two equations, we get:

$$(x + y + z) - (2x - y + z) = 3 - 1$$

$$-x + 2y = 2 \quad \Rightarrow \quad y = x - 1$$

Substituting $y = x - 1$ into one of the original equations to find z:

$$x + (x-1) + z = 3 \quad \Rightarrow \quad 2x + z - 1 = 3 \quad \Rightarrow \quad z = 4 - 2x$$

Thus, the parametric equation of the line of intersection is:

$$(x, y, z) = (t, t-1, 4-2t), \quad t \in \mathbb{R}$$

This describes the intersection of the planes P_1 and P_2. ∎

Lema 5.3.3 If two planes P_1 and P_2 are not parallel, then they always intersect in a line. This line can be found by solving the system of equations representing both planes.

Demostración. The planes $P_1 : A_1 x + B_1 y + C_1 z + D_1 = 0$ and $P_2 : A_2 x + B_2 y + C_2 z + D_2 = 0$ are not parallel if their normal vectors $\vec{n_1} = (A_1, B_1, C_1)$ and $\vec{n_2} = (A_2, B_2, C_2)$ are not proportional. In this case, the system of equations can be solved to obtain a family of solutions that defines a line. The intersection is described using parameters, implying it is a line. ∎

Theorem 5.3.5 The intersection of two planes in $\mathbb{R}^3$ is a line if and only if the planes are not parallel and do not coincide. If the normal vectors of the planes are linearly independent, there is a unique line of intersection.

Demostración. The intersection of two planes in $\mathbb{R}^3$ is a line if the planes are neither parallel nor coincident. This occurs if the normal vectors $\vec{n_1}$ and $\vec{n_2}$ of the planes are linearly independent, meaning there is no scalar $\lambda \neq 0$ such that $\vec{n_2} = \lambda \vec{n_1}$. If the normal vectors are independent, the planes intersect in a unique line. Otherwise, if they are proportional, the planes are either parallel or coincident, and the intersection will be empty or the entire plane. ∎

Corollary 5.3.6 If three planes intersect and are not parallel, their intersection can be either a point or a line. When the three planes are concurrent, they all pass through the same point, which is their intersection.

Demostración. If three planes intersect and are not parallel, there are two possible outcomes: the intersection can be a line if two of the planes are coplanar and the third plane cuts through them, or it can be a point if the three planes are concurrent. The planes are concurrent if they all pass through the same point, in which case this point is the only common intersection of the three planes. ∎

■ **Example 5.17** Consider the planes $P_3 : x - y + 2z = 0$ and $P_4 : 3x + y - z = 5$. To find their intersection, we add and subtract the equations to eliminate variables:

$$x - y + 2z = 0$$

$$3x + y - z = 5$$

Multiplying the first equation by 3 and subtracting the second:

$$3x - 3y + 6z - (3x + y - z) = -5$$

$$-4y + 7z = -5 \quad \Rightarrow \quad y = \frac{7z + 5}{4}$$

We can then describe the intersection using z as a parameter, obtaining the parametric equation of the line of intersection. ∎

5.4 Solved Exercises

> **R** The intersection of planes is a fundamental tool in spatial geometry and linear algebra, as it allows precise descriptions of relationships between different surfaces. Analyzing intersections is also used in solving linear systems and determining the geometry of spatial figures.

Exercise 5.25 Determine the equation of the line of intersection of the planes $2x - y + z = 3$ and $x + 2y - z = 1$.

Exercise 5.26 Verify if the planes $x + y + z = 6$ and $3x - y + 2z = 4$ intersect, and if so, find the parametric equation of the line of intersection.

> **R** The proposed exercises will help the reader better understand how to find and describe the intersection of planes, which is an essential skill in spatial geometry problems and linear systems of equations.

5.4 Solved Exercises

Exercise 5.27 Determine the equation of the line passing through the points $(2,3)$ and $(-1,4)$.

Demostración. To find the equation of the line, we first calculate the slope m using the formula:

$$m = \frac{y_2 - y_1}{x_2 - x_1} = \frac{4-3}{-1-2} = \frac{1}{-3} = -\frac{1}{3}$$

Then, we use the point-slope form of the line equation, taking the point $(2,3)$:

$$y - 3 = -\frac{1}{3}(x - 2)$$

$$y = -\frac{1}{3}x + \frac{2}{3} + 3$$

$$y = -\frac{1}{3}x + \frac{11}{3}$$

∎

Exercise 5.28 Find the equation of the plane passing through the point $(1, -2, 3)$ and with normal vector $\vec{n} = (2, -1, 1)$.

Demostración. The equation of a plane passing through the point (x_0, y_0, z_0) and with normal vector $\vec{n} = (a, b, c)$ is:

$$a(x - x_0) + b(y - y_0) + c(z - z_0) = 0$$

Substitute the values:

$$2(x - 1) - 1(y + 2) + 1(z - 3) = 0$$

$$2x - 2 - y - 2 + z - 3 = 0$$

$$2x - y + z - 7 = 0$$

∎

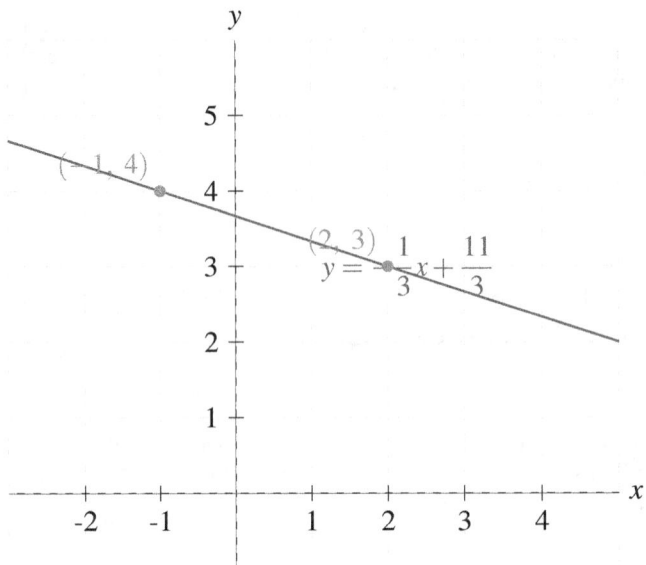

Figura 5.4.1: *Graph of the line passing through the points $(2,3)$ and $(-1,4)$, with equation $y = -\frac{1}{3}x + \frac{11}{3}$.*

Exercise 5.29 Calculate the angle of inclination of the line with slope $m = 2$.

Demostración. The angle of inclination θ of a line is given by:

$$\tan(\theta) = m$$

For $m = 2$:

$$\tan(\theta) = 2$$

$$\theta = \tan^{-1}(2) \approx 63{,}43°$$

∎

Exercise 5.30 Verify if the point $(1,2,3)$ belongs to the plane given by the equation $2x - y + z = 3$.

Demostración. Substitute the point $(1,2,3)$ into the plane equation:

$$2(1) - 2 + 3 = 2 - 2 + 3 = 3$$

Since the equality holds, the point $(1,2,3)$ belongs to the plane.

∎

Exercise 5.31 Determine the parametric equation of the line passing through the point $(0,1,2)$ and parallel to the vector $\vec{v} = (1,2,3)$.

Demostración. The parametric equation of a line passing through the point (x_0, y_0, z_0) and parallel to the vector $\vec{v} = (a,b,c)$ is:

$$x = x_0 + at, \quad y = y_0 + bt, \quad z = z_0 + ct$$

For the point $(0,1,2)$ and the vector $\vec{v} = (1,2,3)$:

$$x = 0 + 1t = t$$

$$y = 1 + 2t$$

$$z = 2 + 3t$$

Thus, the parametric equation is:

$$(x,y,z) = (t, 1+2t, 2+3t)$$

■

These are the solutions to the proposed exercises in Chapter 4. Each problem has been solved step by step to illustrate fundamental concepts related to lines and planes.

5.5 Proposed Exercises

5.5.1 General and Parametric Equations of a Line

Exercise 5.32 Find the parametric equation of the line passing through the points $(1,2,3)$ and $(4,5,6)$. ■

Exercise 5.33 Determine the general equation of the line passing through the point $(3,-1)$ and with a slope of 2. ■

Exercise 5.34 Verify if the point $(2,3)$ belongs to the line $y = 3x+1$. ■

Exercise 5.35 Find the intersection of the lines $y = -x+2$ and $y = 2x-4$. ■

Exercise 5.36 Calculate the distance between the point $(1,2)$ and the line $y = -x+3$. ■

5.5.2 Conditions of Parallelism and Perpendicularity

Exercise 5.37 Determine if the lines $y = 2x+3$ and $y = -\frac{1}{2}x+1$ are perpendicular. ■

Exercise 5.38 Find the slope of the line parallel to $y = -3x+4$ passing through the point $(2,5)$. ■

Exercise 5.39 Verify if the lines $3x-2y+5 = 0$ and $6x-4y+7 = 0$ are parallel. ■

Exercise 5.40 Write the equation of the line perpendicular to $y = -\frac{1}{4}x+1$ and passing through $(0,0)$. ■

Exercise 5.41 Calculate the angle of intersection between the lines $y = x$ and $y = -x$. ■

5.5.3 Equation of the Plane in Space

Exercise 5.42 Find the equation of the plane passing through the points $(1,2,3)$, $(4,5,6)$, and $(7,8,9)$.

Exercise 5.43 Determine if the point $(1,0,-1)$ belongs to the plane $2x-3y+z=4$.

Exercise 5.44 Find the intersection of the plane $x+y+z=6$ with the z-axis.

Exercise 5.45 Calculate the distance between the point $(2,3,4)$ and the plane $3x-y+2z=7$.

Exercise 5.46 Determine the equation of the plane parallel to $x-y+2z=3$ passing through the point $(1,-1,2)$.

6. Equations of Conics

6.1 Circle: Equation and Properties

6.1.1 Equation of the Circle in the Plane

Definition 6.1.1 A **circle** in the plane is the set of all points (x, y) that are at a constant distance r (radius) from a fixed point (h, k), called the **center** of the circle. The standard equation of the circle is:

$$(x-h)^2 + (y-k)^2 = r^2$$

where (h, k) is the center and r is the radius.

■ **Example 6.1** Consider the circle with center at $(3, -2)$ and radius 5. The equation of the circle is:

$$(x-3)^2 + (y+2)^2 = 25$$

We can expand the equation if needed to obtain its general form. ■

Lema 6.1.1 If a point $P = (x_1, y_1)$ belongs to a circle with equation $(x-h)^2 + (y-k)^2 = r^2$, then the distance between the point P and the center (h, k) is equal to the radius r.

Demostración. Given the point $P = (x_1, y_1)$, the distance d between P and the center (h, k) is calculated using the Euclidean distance formula:

$$d = \sqrt{(x_1 - h)^2 + (y_1 - k)^2}$$

Since the point belongs to the circle, this distance must be equal to the radius r. Therefore:

$$\sqrt{(x_1 - h)^2 + (y_1 - k)^2} = r$$

Squaring both sides, we obtain:

$$(x_1 - h)^2 + (y_1 - k)^2 = r^2$$

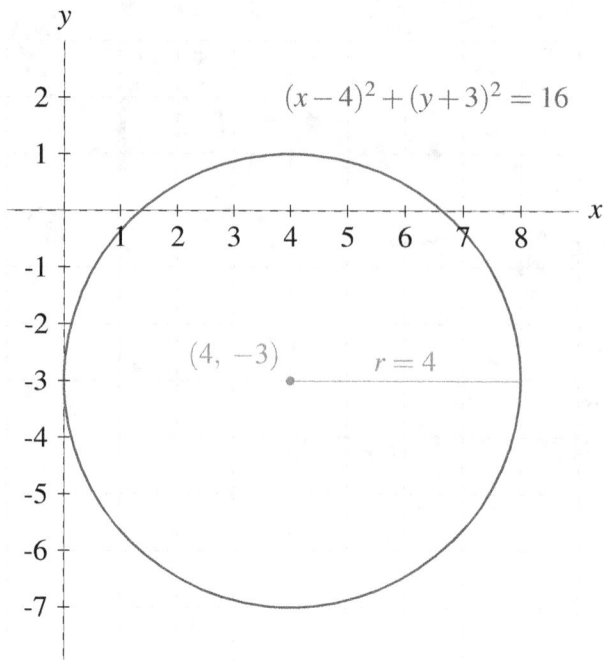

Figura 6.1.1: *Graph of the circle $(x-4)^2 + (y+3)^2 = 16$, with center at $(4, -3)$ and radius $r = 4$.*

which proves that the point P belongs to the circle. ∎

> **Theorem 6.1.1** The general equation of a circle in the plane has the form:
>
> $$x^2 + y^2 + Dx + Ey + F = 0$$
>
> where D, E, and F are constants. To convert this form to the standard form, completing the square is required.

Demostración. Given the general equation $x^2 + y^2 + Dx + Ey + F = 0$, we complete the square to express it in standard form. Group the x and y terms:

$$(x^2 + Dx) + (y^2 + Ey) = -F$$

To complete the square, add and subtract $\left(\frac{D}{2}\right)^2$ and $\left(\frac{E}{2}\right)^2$:

$$\left(x + \frac{D}{2}\right)^2 - \left(\frac{D}{2}\right)^2 + \left(y + \frac{E}{2}\right)^2 - \left(\frac{E}{2}\right)^2 = -F$$

$$\left(x + \frac{D}{2}\right)^2 + \left(y + \frac{E}{2}\right)^2 = \left(\frac{D}{2}\right)^2 + \left(\frac{E}{2}\right)^2 - F$$

This gives the standard equation of the circle, where the center is $\left(-\frac{D}{2}, -\frac{E}{2}\right)$ and the radius is:

$$r = \sqrt{\left(\frac{D}{2}\right)^2 + \left(\frac{E}{2}\right)^2 - F}$$

∎

6.1 Circle: Equation and Properties

Corollary 6.1.2 If the general equation $x^2 + y^2 + Dx + Ey + F = 0$ has a radius $r > 0$, then the equation describes a circle. If $r = 0$, the equation describes a point, and if r is not a real number, no circle exists.

■ **Example 6.2** Consider the equation $x^2 + y^2 - 4x + 6y - 12 = 0$. To find the center and radius, complete the square:

$$(x^2 - 4x) + (y^2 + 6y) = 12$$

$$(x-2)^2 - 4 + (y+3)^2 - 9 = 12$$

$$(x-2)^2 + (y+3)^2 = 25$$

Thus, the center is $(2, -3)$ and the radius is 5. ■

R The standard form of the circle's equation is especially useful for quickly determining the center and radius, while the general form is convenient for algebraic operations. Completing the square is a crucial method for transitioning between these forms.

Exercise 6.1 Determine the center and radius of the circle given by the equation $x^2 + y^2 + 8x - 4y + 7 = 0$.

Exercise 6.2 Verify if the point $P = (1, 2)$ belongs to the circle with equation $(x-3)^2 + (y+1)^2 = 10$.

R The above exercises help the reader practice converting the general form of a circle's equation to the standard form, as well as verifying if a point belongs to a circle. These are fundamental steps in the study of conics and their representation in the plane.

6.1.2 Center and Radius of a Circle

Definition 6.1.2 The **center** of a circle is the fixed point from which all points on the circle are equidistant. This constant distance is called the **radius** of the circle. In its standard form, the equation of a circle is expressed as:

$$(x-h)^2 + (y-k)^2 = r^2$$

where (h, k) represents the center, and r is the radius.

■ **Example 6.3** Consider the equation of a circle $(x-4)^2 + (y+3)^2 = 16$. To identify the center and radius, compare it with the standard form:

$$(x-h)^2 + (y-k)^2 = r^2$$

In this case, the center is $(h, k) = (4, -3)$, and the radius is $r = \sqrt{16} = 4$. ■

Lema 6.1.2 The center and radius of a circle can be obtained by completing the square in the general equation of the circle. The general equation is:

$$x^2 + y^2 + Dx + Ey + F = 0$$

and by completing the square, the standard form can be derived, allowing the center and radius to be identified.

Demostración. To convert the general equation $x^2 + y^2 + Dx + Ey + F = 0$ to the standard form, group the x and y terms and complete the square:

$$(x^2 + Dx) + (y^2 + Ey) = -F$$

Add and subtract $\left(\frac{D}{2}\right)^2$ and $\left(\frac{E}{2}\right)^2$:

$$\left(x + \frac{D}{2}\right)^2 - \left(\frac{D}{2}\right)^2 + \left(y + \frac{E}{2}\right)^2 - \left(\frac{E}{2}\right)^2 = -F$$

$$\left(x + \frac{D}{2}\right)^2 + \left(y + \frac{E}{2}\right)^2 = \left(\frac{D}{2}\right)^2 + \left(\frac{E}{2}\right)^2 - F$$

Thus, the center is $\left(-\frac{D}{2}, -\frac{E}{2}\right)$, and the radius is:

$$r = \sqrt{\left(\frac{D}{2}\right)^2 + \left(\frac{E}{2}\right)^2 - F}$$

∎

Theorem 6.1.3 For any circle in the plane with the general equation $x^2 + y^2 + Dx + Ey + F = 0$, the center and radius can be uniquely determined as long as the value of r is positive. If r is not real, the equation does not represent a circle.

Demostración. By completing the square as shown in the previous lemma, the standard form of the circle is obtained. The radius is positive if $\left(\frac{D}{2}\right)^2 + \left(\frac{E}{2}\right)^2 - F > 0$. If this condition is not satisfied, there is no real value for the radius, implying that no circle exists. ∎

Corollary 6.1.4 If the general equation $x^2 + y^2 + Dx + Ey + F = 0$ has a radius $r = 0$, the circle reduces to a point, and this point is precisely the center of the circle.

■ **Example 6.4** Given the equation $x^2 + y^2 - 6x + 8y + 9 = 0$, find the center and radius by completing the square:

$$(x^2 - 6x) + (y^2 + 8y) = -9$$

$$(x - 3)^2 - 9 + (y + 4)^2 - 16 = -9$$

$$(x - 3)^2 + (y + 4)^2 = 16$$

Thus, the center is $(3, -4)$, and the radius is 4. ■

> (R) It is important to note that converting the general form to the standard form through the method of completing the square is a fundamental technique in analytic geometry. It simplifies the identification of the center and radius of the circle, providing crucial information about its location and size.

6.1 Circle: Equation and Properties

> **Exercise 6.3** Determine the center and radius of the circle with the equation $x^2 + y^2 + 10x - 12y + 20 = 0$.

> **Exercise 6.4** Verify if the point $Q = (0, -2)$ belongs to the circle with center at $(1, -1)$ and radius 2.

(R) The above exercises reinforce the skill of working with circles in different forms and verifying whether a given point lies on the circle, which is essential for a comprehensive understanding of conics.

6.1.3 Applications of the Circle in Real-World Problems

> **Definition 6.1.3** The **applications of the circle in real-world problems** include scenarios where circular trajectories or equidistant calculations from a central point are required. The circle is used in fields such as engineering, physics, architecture, and navigation to describe phenomena involving circular motion or structures.

■ **Example 6.5** In engineering, the circle is used to design circular gears that must mesh perfectly. For example, if a gear with a radius of 10 cm is to be designed, the circumference of this gear is:

$$C = 2\pi r = 2\pi \times 10 = 20\pi \approx 62{,}83 \, \text{cm}$$

This circumference determines the gear's edge, which must match the adjacent gear for smooth and constant motion. ■

Lema 6.1.3 If an object undergoes circular motion with a constant radius r, the length of its path after a full revolution is equal to the circumference, that is, $2\pi r$. This result is fundamental in studying uniform circular motion.

Demostración. Consider an object moving along a circular trajectory of radius r. The distance traveled in a full revolution is equal to the circumference of the circle with that radius. By definition, the circumference is $2\pi r$, so the total path length is exactly that. ■

> **Theorem 6.1.5** The area of the region enclosed by a circle with radius r is:
>
> $$A = \pi r^2$$
>
> This formula is used in various applications, such as calculating the area of circular gardens, the surface of tanks, and more.

Demostración. The formula for the area of the region enclosed by a circle is derived from the concept of integration. Consider a circle with radius r. The region can be thought of as being composed of infinitely thin concentric rings extending from the center to the boundary.
The area of an infinitesimal ring with radius r' and differential thickness dr' is $dA = 2\pi r' \, dr'$. Integrating from the center ($r' = 0$) to the total radius ($r' = r$), we get the total area:

$$A = \int_0^r 2\pi r' \, dr' = \pi r^2$$

Thus, the area of the region enclosed by a circle with radius r is πr^2. ■

Corollary 6.1.6 If the radius of a circle is doubled, the area of the region enclosed by the circle quadruples. That is, if $r \to 2r$, then the new area is $A' = \pi(2r)^2 = 4\pi r^2$.

Demostración. If the original radius of the circle is r, the area is $A = \pi r^2$.
When the radius is doubled, i.e., $r' = 2r$, the new area is calculated as:

$$A' = \pi(r')^2 = \pi(2r)^2 = \pi \cdot 4r^2 = 4\pi r^2$$

Thus, the area quadruples when the radius is doubled. ■

Exercise 6.5 A park has a circular fountain with a radius of 10 meters. Find the length of the fence needed to completely surround the fountain.

Exercise 6.6 A racetrack has a circular shape with a radius of 50 meters. Calculate the area of the track and determine how much surface is needed to cover it if the material costs 3 dollars per square meter.

(R) The above exercises are designed to reinforce the understanding of how the properties of the circle are applied in practical situations, from constructing structures to designing specific areas.

6.2 Parabola: Equation and Applications

6.2.1 Definition and Elements of the Parabola

Definition 6.2.1 A **parabola** is the geometric locus of all points in a plane that are equidistant from a fixed point called the **focus** and a fixed line called the **directrix**. The standard equation of a parabola with its vertex at the origin and axis of symmetry parallel to the y-axis is:

$$y^2 = 4px$$

where p is the distance from the vertex to the focus and from the vertex to the directrix.

■ **Example 6.6** Consider the parabola with the equation $y^2 = 8x$. To determine its elements, we compare it with the standard form:

$$y^2 = 4px$$

In this case, $4p = 8$, so $p = 2$. This means the distance from the vertex to the focus is 2 units, and the directrix is 2 units away from the vertex on the opposite side. The focus of the parabola is $(2,0)$, and the directrix is the line $x = -2$. ■

Lema 6.2.1 The vertex of a parabola is the point of intersection between the parabola's axis of symmetry and the parabola itself. This point is also the closest point on the curve to the focus.

Demostración. By definition, the vertex is the point that lies exactly between the focus and the directrix, equidistant from both. Due to the symmetry of the parabola, the vertex is the point where the axis of symmetry intersects the curve, representing the point of minimal or maximal distance to the axis of the parabola, depending on its orientation. ■

6.2 Parabola: Equation and Applications

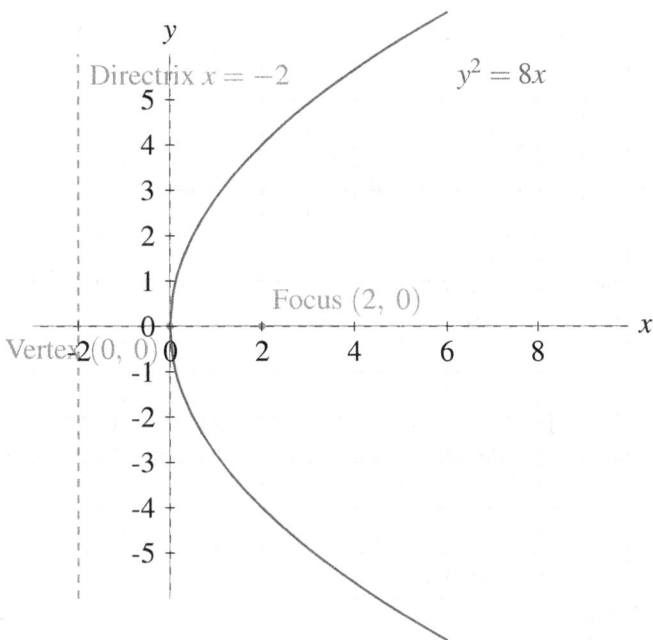

Figura 6.2.1: *Graph of the parabola $y^2 = 8x$, showing the vertex, focus, and directrix.*

Theorem 6.2.1 The general equation of a parabola with its vertex at (h,k) and opening to the right is:

$$(y-k)^2 = 4p(x-h)$$

This equation can be transformed into the standard form to find the focus and directrix.

Demostración. The equation of a parabola with its vertex at (h,k) and opening to the right can be derived from the standard form of a parabola with its vertex at the origin.
For a parabola that opens to the right and has its vertex at the origin, the equation is:

$$y^2 = 4px$$

where p is the distance from the vertex to the focus.
If we translate the vertex to the point (h,k), the equation becomes:

$$(y-k)^2 = 4p(x-h)$$

This equation represents a parabola with its vertex at (h,k), opening to the right and having a focal distance of p. This form allows easy identification of the focus and the directrix. ∎

Corollary 6.2.2 If $p > 0$, the parabola opens to the right or upward, depending on the equation's orientation. If $p < 0$, the parabola opens to the left or downward.

■ **Example 6.7** Given the equation $(y-1)^2 = 12(x+2)$, determine the elements of the parabola. Comparing with the standard form $(y-k)^2 = 4p(x-h)$, we identify $h = -2$, $k = 1$, and $4p = 12$, so $p = 3$. The focus of the parabola is at $(h+p,k) = (1,1)$, and the directrix is the line $x = -5$. ■

(R) The parabola is one of the fundamental conic sections that frequently appear in physical and engineering applications, such as in the design of parabolic antennas, reflectors, and projectile trajectories. Understanding its elements is essential to describe and analyze its behavior in these applications.

> **Exercise 6.7** Find the focus and the equation of the directrix of the parabola with the equation $x^2 = -8y$. What is the direction of the parabola's opening?

> **Exercise 6.8** For the parabola with equation $(x+1)^2 = -12(y-2)$, determine the position of the vertex, the focus, and the equation of the directrix.

> (R) The above exercises will help you familiarize yourself with identifying the fundamental elements of a parabola, such as the focus and the directrix, and deducing its orientation and general form from its equation.

6.2.2 Equation of the Parabola in the Plane

> **Definition 6.2.2** A **parabola** is the set of all points (x,y) in the plane that are equidistant from a fixed point called the **focus** and a fixed line called the **directrix**. The standard equation of a parabola with its vertex at the origin and axis of symmetry parallel to the x-axis is:
>
> $$x^2 = 4py$$
>
> where p is the distance from the vertex to the focus and from the vertex to the directrix. In general, the equation can vary depending on the orientation of the parabola.

■ **Example 6.8** Consider the parabola with equation $x^2 = 12y$. To determine its elements, we compare it with the standard form:

$$x^2 = 4py$$

In this case, $4p = 12$, so $p = 3$. This means the focus is 3 units away from the vertex, in the positive direction of the y-axis. The directrix, on the other hand, is 3 units away from the vertex in the opposite direction, and it is the line $y = -3$. ■

Lema 6.2.2 The distance from the vertex to the focus of a parabola is the same as the distance from the vertex to the directrix. This property defines the nature of the parabola and the symmetry it exhibits with respect to its principal axis.

Demostración. By definition, the parabola is the set of points equidistant from the focus and the directrix. Let V be the vertex of the parabola, F the focus, and D the directrix. The distance from the vertex V to the focus F is p, while the distance from the vertex to the directrix is also p. Since the points on the parabola satisfy this condition of equidistance, the vertex is equidistant from the focus and the directrix. ∎

> **Theorem 6.2.3** The general equation of a parabola with its vertex at the point (h,k) and axis of symmetry parallel to the y-axis can be expressed as:
>
> $$(x-h)^2 = 4p(y-k)$$

6.2 Parabola: Equation and Applications

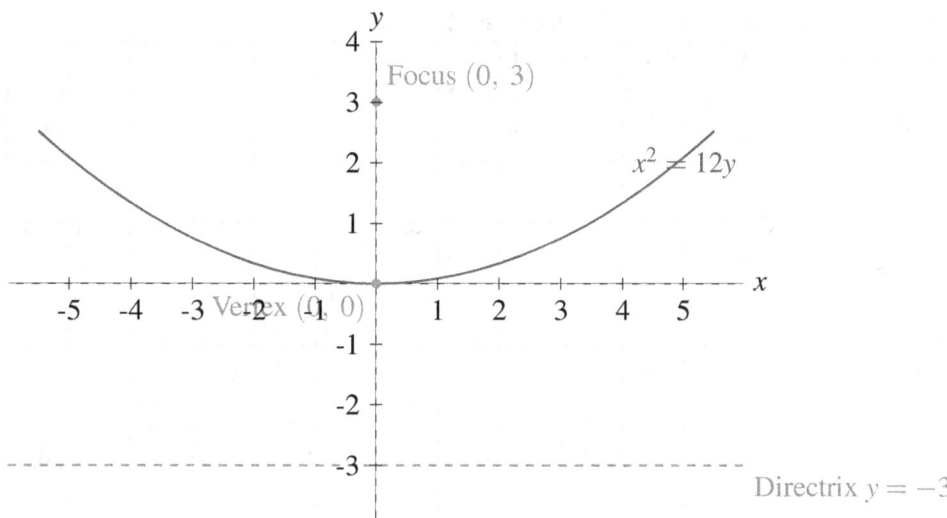

Figura 6.2.2: *Graph of the parabola $x^2 = 12y$, showing the vertex, focus, and directrix.*

where (h, k) is the vertex of the parabola and p is the distance from the vertex to the focus. This form allows the focus and directrix of the parabola to be determined quickly.

Demostración. If the parabola has its vertex at (h, k) and opens upward or downward, we can translate the standard equation $(x-h)^2 = 4p(y-k)$. By doing so, we observe that the focus is at a distance p from the vertex in the direction of the axis of opening, and the directrix is on the opposite side of the vertex at the same distance. ∎

Corollary 6.2.4 If $p > 0$, the parabola opens upward or to the right, depending on the orientation of the axis of symmetry. If $p < 0$, the parabola opens downward or to the left.

■ **Example 6.9** Given the equation $(x+1)^2 = 8(y-2)$, let us determine the elements of the parabola. Comparing with the standard form $(x-h)^2 = 4p(y-k)$, we identify $h = -1$, $k = 2$, and $4p = 8$, which implies $p = 2$. Therefore, the vertex of the parabola is $(-1, 2)$, the focus is at $(h, k+p) = (-1, 4)$, and the directrix is the line $y = 0$. ∎

> Parabolas have numerous applications in real-world problems, such as projectile trajectories, the shape of parabolic antennas, and in architecture to reflect sound or light. Knowing the equation of the parabola allows us to model these phenomena accurately.

Exercise 6.9 Determine the focus and the equation of the directrix of the parabola with the equation $x^2 = -16y$. Additionally, specify the direction of the parabola's opening.

Exercise 6.10 For the parabola with the equation $(y-3)^2 = -12(x+2)$, determine the vertex, the focus, and the equation of the directrix. In which direction does the parabola open?

> These exercises are designed to reinforce the ability to identify the fundamental elements of a parabola and understand how the standard and general equations relate in the analysis of conics.

6.2.3 Applications of the Parabola in Physics and Geometry

Definition 6.2.3 A **parabola** is a curve that has the property of being the geometric locus of points equidistant from a fixed point, called the **focus**, and a fixed line, called the **directrix**. In physics and geometry, parabolas play a fundamental role in phenomena such as reflection and projectile trajectories.

■ **Example 6.10** Consider the motion of an object launched into the air. In the absence of air resistance and under the action of gravity, the trajectory of a projectile is a parabola. For instance, if an object is launched from the origin with an initial velocity $\vec{v}_0$ having horizontal and vertical components, its trajectory equation in the plane can be represented by a parabola. This example demonstrates how parabolas are directly applied to the study of motion in physics. ■

Lema 6.2.3 In a parabolic mirror, any light ray parallel to the axis of symmetry of the parabola is reflected through the focus. This property is used in the construction of **parabolic antennas** and **light reflectors**.

Demostración. By the geometric definition of the parabola, all light rays incident on the surface of a parabolic mirror that are parallel to the axis of symmetry are reflected in such a way that they converge at the focus. This happens because the distance from any point on the parabola to the focus equals the distance from that same point to the directrix, which determines the direction of reflection. ∎

Theorem 6.2.5 The general equation of a parabola that models the trajectory of a projectile, neglecting air resistance, is given by:

$$y = ax^2 + bx + c$$

where a, b, and c are constants depending on the initial velocity and the angle of launch. This equation represents a parabola that opens downward in the context of gravity.

Corollary 6.2.6 The maximum height of a projectile, in the equation of the form $y = ax^2 + bx + c$, is reached at the vertex of the parabola. The x-coordinate of the vertex is given by:

$$x = -\frac{b}{2a}$$

and the maximum height can be calculated by substituting this value into the equation of the parabola.

■ **Example 6.11** Consider a projectile launched with an initial velocity of 50 m/s at an angle of 30° with respect to the ground. The trajectory equation can be written as:

$$y = -4{,}9x^2 + 28{,}87x$$

where components of the initial velocity and the acceleration due to gravity are used. The maximum height can be calculated by substituting the x-coordinate of the vertex into the equation. ■

> ® The applications of the parabola in physics also include the construction of bridges and reflectors. The geometry of the parabola ensures that loads on a parabolic bridge are distributed efficiently, while its reflective property is applied in antennas and solar energy systems.

6.3 Ellipse and Hyperbola: Definition and Characteristics

> **Exercise 6.11** A parabolic antenna has the shape of a parabola with its focus located 4 meters from the vertex. Find the equation of the parabola that describes the shape of the antenna, assuming the vertex is at the origin and the parabola opens upward.

> **Exercise 6.12** In a physics experiment, a projectile is launched from the ground with an initial velocity that produces a parabolic trajectory. If the equation of the trajectory is $y = -5x^2 + 20x$, determine the maximum height reached by the projectile and the horizontal distance it travels before hitting the ground again.

> (R) These exercises help students understand how physical phenomena are modeled using parabolas, from the reflection of light in a parabolic mirror to the trajectory of a moving projectile.

6.3 Ellipse and Hyperbola: Definition and Characteristics

6.3.1 Definition and Elements of the Ellipse

Definition 6.3.1 An **ellipse** is the set of all points in a plane for which the sum of the distances to two fixed points, called **foci**, is constant. The standard equation of an ellipse centered at the origin with axes aligned with the coordinate axes is:

$$\frac{x^2}{a^2} + \frac{y^2}{b^2} = 1$$

where a and b are the lengths of the semi-major and semi-minor axes, respectively. If $a > b$, the ellipse is elongated along the x-axis, whereas if $b > a$, it is elongated along the y-axis.

■ **Example 6.12** Consider the ellipse with the equation:

$$\frac{x^2}{9} + \frac{y^2}{4} = 1$$

Here, $a^2 = 9$, so $a = 3$, and $b^2 = 4$, which implies $b = 2$. This means the length of the semi-major axis is 3 units, and the length of the semi-minor axis is 2 units. The foci are located along the x-axis at a distance of $\sqrt{a^2 - b^2} = \sqrt{5}$ from the center. Therefore, the foci are at the points $(\pm\sqrt{5}, 0)$. ■

Lema 6.3.1 The distance between the two foci of an ellipse is $2c$, where $c = \sqrt{a^2 - b^2}$ if $a > b$, or $c = \sqrt{b^2 - a^2}$ if $b > a$. This distance is crucial to defining the shape of the ellipse and its geometric properties.

> **Theorem 6.3.1** The eccentricity e of an ellipse is defined as:
>
> $$e = \frac{c}{a}$$
>
> where c is the distance from the center to each of the foci, and a is the length of the semi-major axis. The eccentricity of an ellipse lies in the range $0 < e < 1$. If $e = 0$, the ellipse becomes a circle, while as e approaches 1, the ellipse becomes more elongated.

Demostración. By definition, $c = \sqrt{a^2 - b^2}$. The eccentricity is defined as the ratio of the focal distance c to the length of the semi-major axis a. Since $a > b > 0$, it follows that $0 < c < a$, implying $0 < e < 1$. If $b = a$, then $c = 0$, and the ellipse becomes a circle. ∎

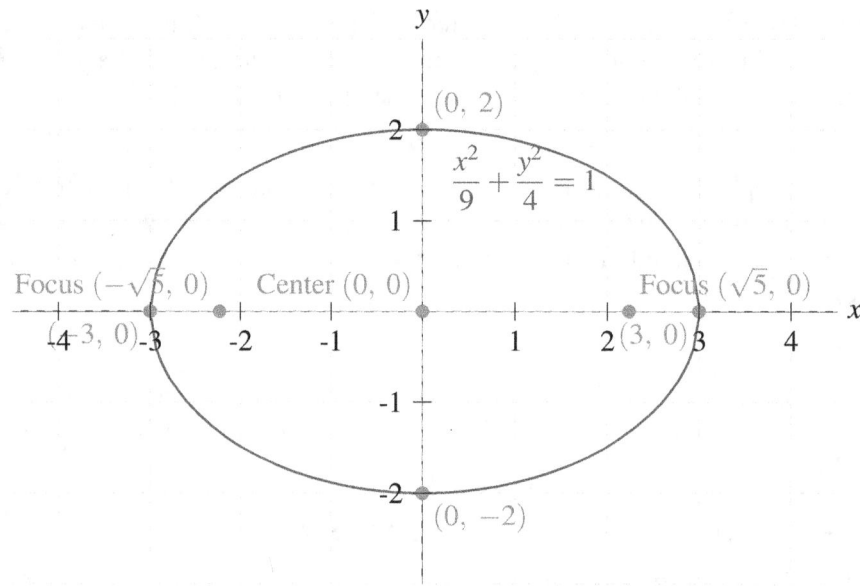

Figura 6.3.1: *Graph of the ellipse* $\dfrac{x^2}{9} + \dfrac{y^2}{4} = 1$, *showing the center, vertices, and foci.*

Corollary 6.3.2 If $a = b$, the ellipse reduces to a **circle** with radius $r = a = b$, and the eccentricity is $e = 0$. This implies that a circle is a special case of an ellipse with equal lengths for both semi-axes.

■ **Example 6.13** Suppose we have an ellipse with a semi-major axis $a = 5$ and a semi-minor axis $b = 3$. The focal distance is:

$$c = \sqrt{a^2 - b^2} = \sqrt{25 - 9} = \sqrt{16} = 4$$

Thus, the foci are at the points $(\pm 4, 0)$. The eccentricity of the ellipse is:

$$e = \frac{c}{a} = \frac{4}{5} = 0{,}8$$

This shows that the ellipse is relatively elongated, with the foci located near the ends of the semi-major axis. ■

> The eccentricity of an ellipse determines how "flattened.°r roundedït is. When e approaches 0, the ellipse resembles a circle, while as e approaches 1, the ellipse becomes increasingly elongated. This characteristic has important applications in physics, such as in the description of planetary orbits.

Exercise 6.13 Given the equation of an ellipse $\dfrac{x^2}{16} + \dfrac{y^2}{9} = 1$, find the lengths of the semi-major and semi-minor axes, and determine the location of the foci.

Exercise 6.14 Calculate the eccentricity of the ellipse whose equation is $\dfrac{x^2}{25} + \dfrac{y^2}{9} = 1$. Determine whether this ellipse is more elongated or closer to the shape of a circle.

6.3 Ellipse and Hyperbola: Definition and Characteristics

 These exercises help reinforce the understanding of the fundamental elements of the ellipse, such as the semi-axes, foci, and eccentricity, which are key to describing its shape and geometric properties.

6.3.2 Definition and Elements of the Hyperbola

Definition 6.3.2 A **hyperbola** is the set of all points in the plane such that the absolute difference of their distances to two fixed points, called **foci**, is constant. The standard equation of a hyperbola centered at the origin with axes aligned with the coordinate axes is:

$$\frac{x^2}{a^2} - \frac{y^2}{b^2} = 1$$

where a and b are constants that determine the shape of the hyperbola. If $a > b$, the hyperbola opens along the x-axis, and if $b > a$, it opens along the y-axis.

■ **Example 6.14** Consider the hyperbola with the equation:

$$\frac{x^2}{16} - \frac{y^2}{9} = 1$$

In this case, $a^2 = 16$, so $a = 4$, and $b^2 = 9$, which implies $b = 3$. The foci of the hyperbola are located along the x-axis at a distance of $c = \sqrt{a^2 + b^2} = \sqrt{16+9} = 5$ from the center. Therefore, the foci are at the points $(\pm 5, 0)$. ■

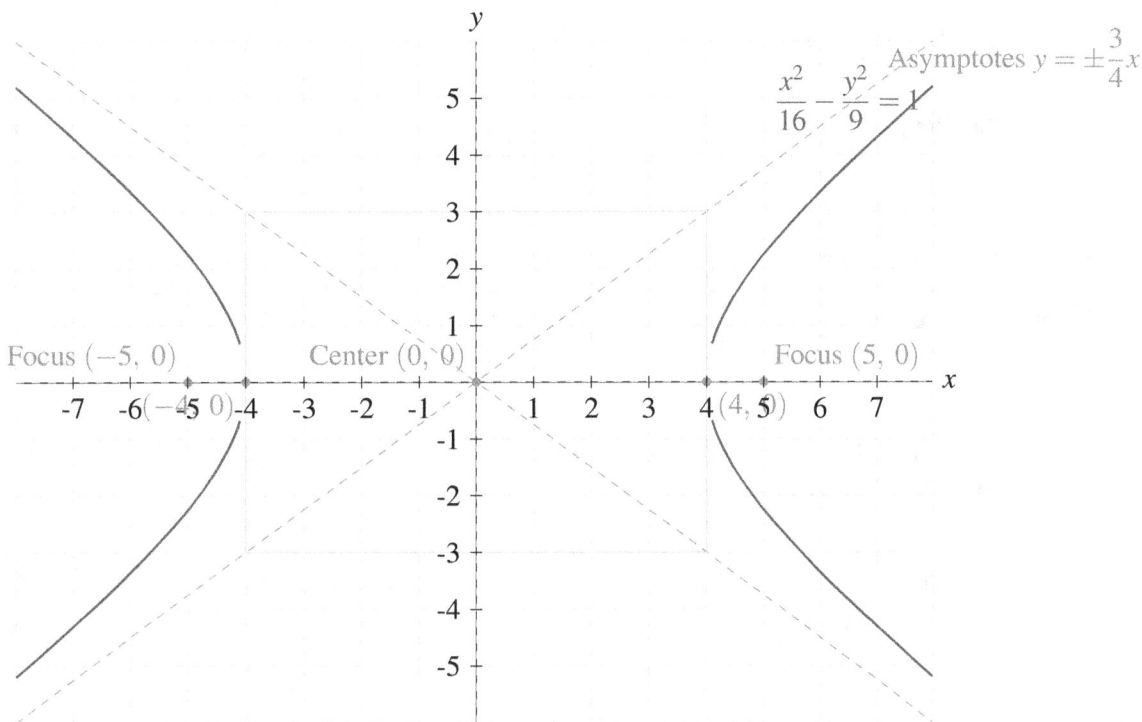

Figura 6.3.2: *Graph of the hyperbola $\frac{x^2}{16} - \frac{y^2}{9} = 1$, showing the center, vertices, foci, and asymptotes.*

Lema 6.3.2 The distance between the two foci of a hyperbola is $2c$, where $c = \sqrt{a^2 + b^2}$. This distance is crucial for defining the shape and the opening of the hyperbola.

Demostración. By definition, the distance between the foci of a hyperbola is twice the distance from the center to each focus. Since $c = \sqrt{a^2 + b^2}$, the total distance between the foci is $2c$. This distance reflects the extent of the hyperbola along the axis containing the foci. ∎

> **Theorem 6.3.3** The **eccentricity** e of a hyperbola is defined as:
>
> $$e = \frac{c}{a}$$
>
> where $c = \sqrt{a^2 + b^2}$. The eccentricity of a hyperbola is always greater than 1. The greater the value of e, the more "flattened" the hyperbola becomes.

> **Corollary 6.3.4** If $e > 1$, the curve is a hyperbola. This result confirms that the eccentricity of a hyperbola always exceeds that of an ellipse ($e < 1$) and a circle ($e = 0$), explaining the open shape of the hyperbola.

■ **Example 6.15** Suppose we have a hyperbola with a transverse axis $a = 6$ and a conjugate axis $b = 8$. The focal distance is:

$$c = \sqrt{a^2 + b^2} = \sqrt{36 + 64} = 10$$

Thus, the foci are at the points $(\pm 10, 0)$. The eccentricity of the hyperbola is:

$$e = \frac{c}{a} = \frac{10}{6} = \frac{5}{3} \approx 1{,}67$$

This shows that the hyperbola has an eccentricity greater than 1, as expected, and the foci are far from the center, determining the curve's openness. ■

> (R) The hyperbola has many applications in both mathematics and physics. It is used to describe phenomena such as the trajectories of objects under certain repulsive forces and in GPS navigation systems. Additionally, the reflective properties of the hyperbola are applied in the design of telescopes and antennas.

> **Exercise 6.15** Given the equation of a hyperbola $\frac{x^2}{25} - \frac{y^2}{9} = 1$, find the lengths of the transverse and conjugate axes, and determine the location of the foci.

> **Exercise 6.16** Calculate the eccentricity of the hyperbola whose equation is $\frac{y^2}{36} - \frac{x^2}{16} = 1$. Determine whether this hyperbola is very open or closer to being an ellipse.

> (R) These exercises are designed to help the reader understand how to identify the fundamental elements of the hyperbola, such as the axes, foci, and eccentricity, and how these properties influence the shape and applications of the curve.

6.3.3 Applications of the Ellipse and Hyperbola

Definition 6.3.3 **Ellipses** and **hyperbolas** are conic sections with various real-life and scientific applications. The ellipse is defined as the set of points whose sum of distances to two foci is constant, while the hyperbola is the set of points whose difference of distances to two foci is constant. Both have applications ranging from engineering to physics and astronomy.

■ **Example 6.16** A practical application of the **ellipse** is found in reflectors and acoustic systems. In an elliptical reflector, any ray of light or sound that hits the surface and passes through one focus will reflect and pass through the other focus. This property is used in concert halls to improve acoustics and in antenna systems to focus electromagnetic waves. ■

■ **Example 6.17** The **hyperbola** has important applications in navigation and physics. In particular, the GPS system uses hyperbolas to determine the exact position of an object on Earth. Satellite signals generate hyperbolic curves, and by knowing the differences in arrival times from at least three satellites, the object's location can be precisely determined. ■

Lema 6.3.3 The sum of the distances from any point on an ellipse to its two foci is always constant and equal to the length of the major axis of the ellipse. This property is fundamental for explaining how sound and light waves reflect within elliptical structures, concentrating these waves at the second focus.

Theorem 6.3.5 The **hyperbola** is used to model escape trajectories in physics. If an object has enough energy to escape the gravitational pull of a planet, its trajectory will be a hyperbola with respect to the planet. This property is used in space missions to plan routes that allow spacecraft to escape Earth's gravitational field.

Corollary 6.3.6 If a body has kinetic energy exceeding the required escape energy, its trajectory around a celestial body is no longer an ellipse but a hyperbola. This result explains why the trajectories of space probes, as they leave the solar system, become hyperbolic.

(R) The applications of the ellipse and hyperbola are fundamental to modern technology. The ellipse helps concentrate energy in applications such as parabolic antennas and reflection systems, while the hyperbola enables efficient navigation routes in satellite systems and escape trajectories in astronomy.

Exercise 6.17 A concert hall has its ceiling designed in the shape of an ellipse with one focus located at the stage. If a point on the ceiling is located at the other focus of the ellipse, demonstrate that the speaker's voice reflects off the ceiling and reaches the audience. Explain why the property of the ellipse is useful in this context. ■

Exercise 6.18 The GPS system uses hyperbolas to determine the location of a receiver on Earth. If signals are received from two satellites, determine how the time differences can be used to define a hyperbola that helps locate the receiver. ■

(R) These exercises are designed to help the reader understand how the properties of conic sections are applied in practice, from architectural design to satellite navigation.

6.4 Solved Exercises

> **Exercise 6.19** Find the center and radius of the circle given by the equation $x^2 + y^2 - 6x + 4y - 3 = 0$.

Solution:
First, we complete the square for the equation:

$$x^2 - 6x + y^2 + 4y = 3$$

$$(x^2 - 6x) + (y^2 + 4y) = 3$$

Complete the square for each variable:

$$(x^2 - 6x + 9) + (y^2 + 4y + 4) = 3 + 9 + 4$$

$$(x-3)^2 + (y+2)^2 = 16$$

This is the equation of a circle with center $(3, -2)$ and radius $r = \sqrt{16} = 4$.
Center: $(3, -2)$
Radius: 4

> **Exercise 6.20** Determine the vertex and focus of the parabola $y^2 = 8x$.

Solution:
The equation of the parabola is of the form $y^2 = 4px$, where p is the distance from the vertex to the focus.
Comparing with $y^2 = 8x$, we have $4p = 8 \Rightarrow p = 2$.
The vertex of the parabola is $(0,0)$, and the focus is at a distance of p units from the vertex along the positive x-axis. Therefore, the focus is $(2,0)$.
Vertex: $(0,0)$
Focus: $(2,0)$

> **Exercise 6.21** Calculate the foci and eccentricity of the ellipse $\frac{x^2}{9} + \frac{y^2}{4} = 1$.

Solution:
The equation of the ellipse is of the form $\frac{x^2}{a^2} + \frac{y^2}{b^2} = 1$, where $a > b$.
Here, $a^2 = 9 \Rightarrow a = 3$ and $b^2 = 4 \Rightarrow b = 2$.
The focal distance c is calculated as:

$$c = \sqrt{a^2 - b^2} = \sqrt{9 - 4} = \sqrt{5}$$

Thus, the foci are at $(\pm\sqrt{5}, 0)$.
The eccentricity e is calculated as:

$$e = \frac{c}{a} = \frac{\sqrt{5}}{3}$$

Foci: $(\pm\sqrt{5}, 0)$
Eccentricity: $\frac{\sqrt{5}}{3}$

6.4 Solved Exercises

Exercise 6.22 Find the equation of the hyperbola with foci at $(\pm 5, 0)$ and vertices at $(\pm 3, 0)$.

Solution:
The equation of a hyperbola with center at the origin and a horizontal transverse axis is of the form:

$$\frac{x^2}{a^2} - \frac{y^2}{b^2} = 1$$

The vertices are at $(\pm 3, 0)$, so $a = 3 \Rightarrow a^2 = 9$.
The foci are at $(\pm 5, 0)$, so $c = 5$. Using the relationship $c^2 = a^2 + b^2$, we find b:

$$5^2 = 3^2 + b^2$$

$$25 = 9 + b^2$$

$$b^2 = 16 \Rightarrow b = 4$$

Thus, the equation of the hyperbola is:

$$\frac{x^2}{9} - \frac{y^2}{16} = 1$$

Exercise 6.23 Graph the parabola $y = x^2 - 4x + 3$ and find the coordinates of its vertex.

Solution:
First, complete the square to find the vertex coordinates:

$$y = x^2 - 4x + 3$$

$$y = (x^2 - 4x + 4) - 4 + 3$$

$$y = (x - 2)^2 - 1$$

The vertex form of the parabola is $y = (x - 2)^2 - 1$, indicating that the vertex is at $(2, -1)$.
Vertex: $(2, -1)$
The graph is a parabola opening upwards with the vertex at $(2, -1)$.

6.5 Proposed Exercises

6.5.1 Circle: Equation and Properties

Exercise 6.24 Determine the equation of the circle with center $(3, -2)$ and radius 5.

Exercise 6.25 Find the radius of the circle given by the equation $x^2 + y^2 - 6x + 8y + 9 = 0$.

Exercise 6.26 Graph the circle with the equation $(x-1)^2 + (y+2)^2 = 16$.

Exercise 6.27 Find the circumference of the circle with radius 7.

Exercise 6.28 Determine the area of the circle with the equation $(x+3)^2 + (y-4)^2 = 25$.

6.5.2 Parabola: Equation and Applications

Exercise 6.29 Find the equation of the parabola with vertex at $(0,0)$ and focus at $(0,3)$.

Exercise 6.30 Determine the focus and directrix of the parabola $y^2 = 8x$.

Exercise 6.31 Find the equation of the parabola passing through the points $(0,0)$, $(1,1)$, and $(2,4)$.

Exercise 6.32 Graph the parabola $y = 2x^2 + 3x - 5$.

Exercise 6.33 Calculate the length of the latus rectum of the parabola $y^2 = 4x$.

6.5.3 Ellipse and Hyperbola: Definitions and Characteristics

Exercise 6.34 Determine the equation of the ellipse with foci at $(\pm 3, 0)$ and a major axis of 10.

Exercise 6.35 Find the asymptotes of the hyperbola $9x^2 - 16y^2 = 144$.

Exercise 6.36 Graph the ellipse given by $\frac{x^2}{25} + \frac{y^2}{9} = 1$.

Exercise 6.37 Determine the foci of the hyperbola with the equation $x^2 - y^2 = 1$.

Exercise 6.38 Calculate the eccentricity of the ellipse $\frac{x^2}{36} + \frac{y^2}{16} = 1$.

7. Functions: Domain, Range, and Operations

7.1 Domain and Range of Real Functions

7.1.1 Definition of Domain and Range

Definition 7.1.1 The **domain** of a function is the set of all possible values that can be assigned to the independent variable such that the function is defined. The **range**, on the other hand, is the set of all possible values the function can take as output. In other words, the domain represents the possible input values, and the range represents the possible output values.

■ **Example 7.1** Consider the function $f(x) = \sqrt{x-1}$. For the function to be defined, the value under the square root must be non-negative, i.e., $x - 1 \geq 0$. Thus, the **domain** of the function is $[1, \infty)$. The **range** of the function consists of all possible values of $f(x)$, which in this case is $[0, \infty)$.
■

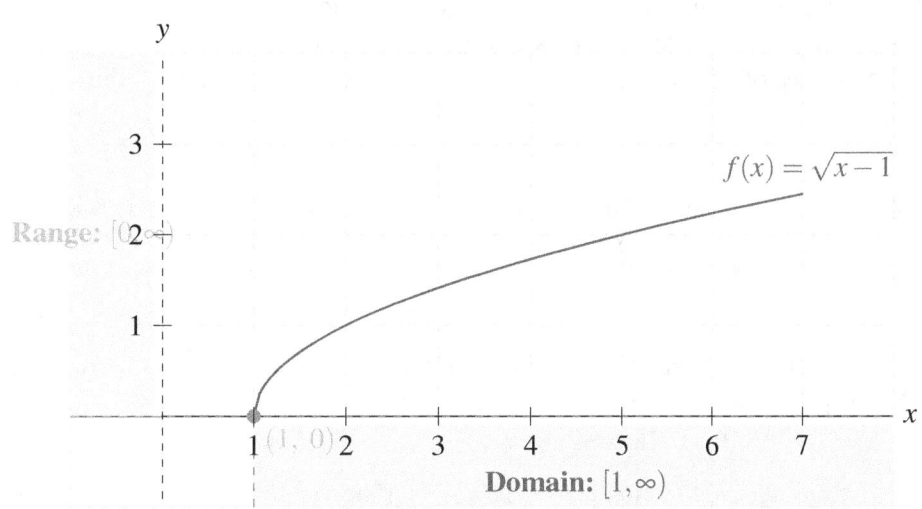

Figura 7.1.1: *Graph of the function* $f(x) = \sqrt{x-1}$, *showing the domain* $[1, \infty)$ *and the range* $[0, \infty)$.

Capítulo 7. Functions: Domain, Range, and Operations

Lema 7.1.1 If a function $f : A \to B$ is injective, then each value in the range is associated with exactly one value in the domain. This implies that no two distinct values in the domain share the same image.

Demostración. Let $f : A \to B$ be an injective function. This means that if $f(x_1) = f(x_2)$, then necessarily $x_1 = x_2$. Therefore, no two distinct values x_1 and x_2 in the domain can correspond to the same value in the range, demonstrating that each value in the range is uniquely associated with one value in the domain. ∎

> **Theorem 7.1.1** The range of a composite function $g(f(x))$ is contained within the range of the outer function g, and it is determined by the set of values the function g can achieve when evaluating the elements of the range of the inner function f.

Demostración. To demonstrate this theorem, consider the composite function $g(f(x))$, where $f : A \to B$ and $g : B \to C$. The range of the composite function depends on the values that g can take when evaluating elements within the range of f.

The range of f is contained in B, so the values of $f(x)$ will lie within the set B. When we evaluate the function g on the values of $f(x)$, these results will lie within the set C, which is the range of the function g.

Thus, the range of the composite function $g(f(x))$ is contained within the range of g, as it depends on the values achievable by g when its input comes from the range of f. ∎

Corollary 7.1.2 If $f : A \to B$ and $g : B \to C$ are two functions where f is surjective and g is injective, then the composite function $g \circ f$ is also surjective.

Demostración. To show that the composite function $g \circ f$ is surjective, we need to prove that for every element $c \in C$, there exists at least one element $a \in A$ such that $(g \circ f)(a) = c$.

Since f is surjective, for every element $b \in B$, there exists an $a \in A$ such that $f(a) = b$. Additionally, since g is injective, for every $c \in C$, there exists a unique $b \in B$ such that $g(b) = c$.

By the surjectivity of f, we can find an $a \in A$ such that $f(a) = b$. Applying g to b, we obtain $g(f(a)) = c$. Therefore, for any $c \in C$, there exists an $a \in A$ such that $(g \circ f)(a) = c$, proving that $g \circ f$ is surjective. ∎

■ **Example 7.2** Consider the functions $f(x) = x^2$ with domain $[0, \infty)$ and $g(x) = x + 1$. The composite function $g(f(x)) = x^2 + 1$ has a domain equal to that of f, i.e., $[0, \infty)$, and a range of $[1, \infty)$, since the values of $f(x)$ are always non-negative, and adding 1 yields values greater than or equal to 1. ∎

> (R) Understanding the concept of domain and range is fundamental for function analysis, as it helps identify the valid input values and possible outputs of a function. This is particularly important when working with composite functions and function transformations.

Exercise 7.1 Determine the domain and range of the function $f(x) = \frac{1}{x-2}$. Also, indicate if the function has any discontinuities and where they occur.

Exercise 7.2 Consider the function $g(x) = \ln(x - 3)$. Find the domain and range of the function, and explain why the function is undefined for certain values of x.

7.1 Domain and Range of Real Functions

> (R) These exercises help reinforce the reader's understanding of how to determine the domain and range of a function, which is crucial for analyzing the properties and behavior of functions.

7.1.2 Determination of the Domain of Rational and Radical Functions

Definition 7.1.2 The **domain** of a rational function is the set of all possible values of the independent variable for which the function is defined. In rational functions, the domain is determined by excluding the values that make the denominator equal to zero. For **radical** functions with an even index, the domain is determined by finding the values that make the radicand greater than or equal to zero, ensuring that the expression is defined in the set of real numbers.

■ **Example 7.3** Consider the rational function $f(x) = \frac{x+1}{x-3}$. For the function to be defined, the denominator must not be zero, i.e., $x - 3 \neq 0$. Thus, $x \neq 3$, and the domain of $f(x)$ is $(-\infty, 3) \cup (3, \infty)$. ■

■ **Example 7.4** Now consider the radical function $g(x) = \sqrt{x-2}$. The radicand must be greater than or equal to zero, i.e., $x - 2 \geq 0$, which implies $x \geq 2$. Thus, the domain of $g(x)$ is $[2, \infty)$. ■

Lema 7.1.2 The domain of a rational function $f(x) = \frac{p(x)}{q(x)}$ is given by all values of x where the denominator $q(x) \neq 0$. This condition ensures that the function is defined for all possible values of the independent variable, except at points of discontinuity where the denominator is zero.

Theorem 7.1.3 Let $f(x)$ be a function composed of a rational function and a radical function, such as $f(x) = \frac{\sqrt{x+1}}{x-2}$. The domain of $f(x)$ is determined by resolving both the restrictions of the denominator and those of the radicand. In this case, the radicand $x + 1 \geq 0$ implies $x \geq -1$, and the denominator $x - 2 \neq 0$ implies $x \neq 2$. Therefore, the domain is $[-1, 2) \cup (2, \infty)$.

Corollary 7.1.4 If a rational function has a denominator that includes a quadratic expression with no real roots, then the domain is the entire set of real numbers. For example, for the function $h(x) = \frac{1}{x^2+1}$, the denominator is never zero for any real value of x, implying that the domain is $(-\infty, \infty)$.

> (R) The domain of a rational or radical function is fundamental for understanding the properties of the function, such as its continuity and points of discontinuity. It is important to carefully analyze the denominator and radicand to correctly determine the set of allowable values.

Exercise 7.3 Determine the domain of the rational function $f(x) = \frac{2x+3}{x^2-4}$. Also, indicate whether the function has any points of discontinuity, and if so, at which values they occur. ■

Exercise 7.4 Find the domain of the radical function $g(x) = \sqrt{5-x}$ and explain why the function is not defined for certain values of x. ■

> (R) These exercises allow the reader to apply the concepts of domain to rational and radical functions, identifying restrictions and points of discontinuity, which are key to understanding the behavior of these functions.

7.1.3 Graphical Representation of Domain and Range

Definition 7.1.3 The **graphical representation of the domain** of a function involves showing all allowable values for the independent variable on the horizontal axis (x-axis), while the **graphical representation of the range** involves showing all possible output values of the function on the vertical axis (y-axis). These representations provide a clear visualization of the restrictions and behavior of the function.

■ **Example 7.5** Consider the function $f(x) = \sqrt{x-1}$. The **domain** of the function is $x \geq 1$, which is graphically represented as a continuous line on the x-axis starting at $x = 1$ and extending to the right. The **range** of the function is $y \geq 0$, represented as a continuous line on the y-axis starting at $y = 0$ and extending upward. ■

Lema 7.1.3 If a function is **continuous** and has a limited domain, then the range will also be limited and determined by the extreme values of the domain. This implies that by graphically representing the domain, a limited corresponding range can be deduced.

Demostración. Let $f : [a,b] \to \mathbb{R}$ be a continuous function defined on the interval $[a,b]$. By the extreme value theorem, f will achieve both a maximum and a minimum value on the interval, limiting the range of the function to these values. This means that the graphical representation of the range will be bounded by these extremes. ∎

Theorem 7.1.5 The graphical representation of the domain and range of a composite function $g(f(x))$ can be determined by first evaluating the domain of the inner function $f(x)$ and then applying the result to the outer function g. This allows for the graphical representation of the possible values for $g(f(x))$.

Corollary 7.1.6 If the inner function of a composite function has a limited domain, the range of the composite function will also be limited by the range of the outer function evaluated over the limited domain. This result simplifies the graphical determination of the range of a composite function.

■ **Example 7.6** Consider the functions $f(x) = x^2$ and $g(x) = \sqrt{x}$. The composite function $g(f(x)) = \sqrt{x^2}$ has a **domain** equal to all $\mathbb{R}$, while the **range** is $[0, \infty)$, as the value of $\sqrt{x^2}$ is always greater than or equal to zero. The graphical representation of the domain is the entire real line on the x-axis, while the representation of the range is the continuous line on the y-axis starting from zero. ■

> (R) The graphical representation of the domain and range not only helps in understanding the definition of a function but also in predicting its behavior. In the case of composite or more complex functions, graphical representation becomes an essential tool for analyzing fundamental characteristics.

Exercise 7.5 Determine the domain and range of the rational function $f(x) = \frac{1}{x-2}$ and describe how they would be graphically represented on the real line. ■

Exercise 7.6 Find the domain and range of the function $g(x) = \ln(x+3)$. Describe how these sets would be graphically represented. ■

> (R) These exercises aim to help the reader visualize and understand how to graphically represent the domain and range of both rational and logarithmic functions, fundamental aspects in functional analysis.

7.2 Operations with Functions (Addition, Subtraction, Multiplication, Division)

7.2.1 Addition and Subtraction of Functions

Definition 7.2.1 Given two functions $f : A \to \mathbb{R}$ and $g : A \to \mathbb{R}$ defined on the same domain A, the **addition** $(f+g)(x)$ and the **subtraction** $(f-g)(x)$ of the functions are defined as follows for all $x \in A$:

$$(f+g)(x) = f(x) + g(x)$$

$$(f-g)(x) = f(x) - g(x)$$

The domain of both operations is the set A, which is the intersection of the domains of f and g.

■ **Example 7.7** Consider the functions $f(x) = 2x+3$ and $g(x) = x^2 - 1$. The **sum** of f and g is obtained by adding both expressions:

$$(f+g)(x) = (2x+3) + (x^2 - 1) = x^2 + 2x + 2$$

The **difference** is calculated by subtracting g from f:

$$(f-g)(x) = (2x+3) - (x^2 - 1) = -x^2 + 2x + 4$$

The domain of both functions is the set of all real numbers, i.e., $(-\infty, \infty)$. ■

Lema 7.2.1 The addition and subtraction of functions are closed operations in the set of continuous functions defined on an interval A. That is, if f and g are continuous on A, then $f+g$ and $f-g$ are also continuous on A.

Demostración. Let f and g be continuous functions on the interval A. By the definition of continuity, for all $x \in A$, $f(x)$ and $g(x)$ are continuous. The sum and difference of continuous functions are also continuous because the sum and difference of limits exist and are finite. This implies that $(f+g)(x)$ and $(f-g)(x)$ are continuous on A. ∎

> **Theorem 7.2.1** The derivative of the sum of two functions is equal to the sum of their derivatives. If f and g are differentiable functions, then:
>
> $$(f+g)'(x) = f'(x) + g'(x)$$
>
> This result also applies to the subtraction of functions.

Corollary 7.2.2 If f and g are differentiable functions, then the derivative of the difference of f and g is equal to the difference of their derivatives:

$$(f-g)'(x) = f'(x) - g'(x)$$

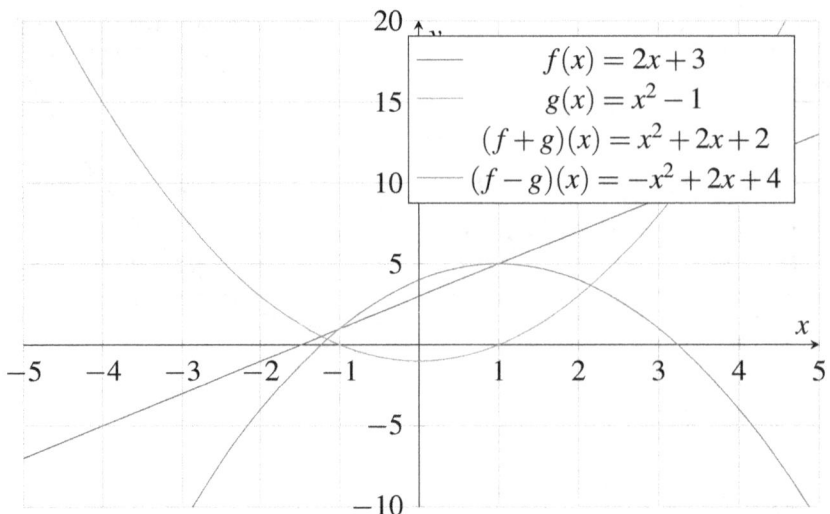

Figura 7.2.1: *Graphs of $f(x)$, $g(x)$, $(f+g)(x)$, and $(f-g)(x)$.*

This corollary is a direct consequence of the linearity of the derivative and is fundamental for calculating the derivative of more complex expressions.

■ **Example 7.8** Consider the functions $f(x) = x^3$ and $g(x) = x^2$. The derivative of the **sum** of f and g is:

$$(f+g)'(x) = f'(x) + g'(x) = 3x^2 + 2x$$

While the derivative of the **difference** of f and g is:

$$(f-g)'(x) = f'(x) - g'(x) = 3x^2 - 2x$$

This example shows how the differentiation rules are applied to the operations of addition and subtraction of functions. ■

> (R) The operations of addition and subtraction of functions allow constructing new functions from existing ones while preserving important properties such as continuity and differentiability. These properties are fundamental in mathematical analysis and problem-solving.

Exercise 7.7 Given the functions $f(x) = \ln(x)$ and $g(x) = \sqrt{x}$, find the expressions for $(f+g)(x)$ and $(f-g)(x)$. Determine the domain of each function. ■

Exercise 7.8 Consider the functions $f(x) = e^x$ and $g(x) = x^3$. Calculate the derivative of $(f+g)(x)$ and the derivative of $(f-g)(x)$. What do you observe about the linearity of the derivative? ■

> (R) These exercises aim to help the reader practice applying the addition and subtraction of functions, as well as verifying the differentiation properties, essential concepts in functional analysis and calculus.

7.2.2 Multiplication and Division of Functions

Definition 7.2.2 The **multiplication** of two functions f and g, defined on a common set A, is denoted as $(f \cdot g)(x) = f(x) \cdot g(x)$ for all $x \in A$. The **division** of two functions f and g is defined as $\left(\frac{f}{g}\right)(x) = \frac{f(x)}{g(x)}$, with the condition that $g(x) \neq 0$ for all $x \in A$. The domain of both operations is the intersection of the domains of f and g, excluding the values where $g(x) = 0$ in the case of division.

■ **Example 7.9** Consider the functions $f(x) = x^2$ and $g(x) = x - 1$. The **multiplication** of f and g is expressed as:

$$(f \cdot g)(x) = x^2 \cdot (x-1) = x^3 - x^2$$

For the **division**, we have:

$$\left(\frac{f}{g}\right)(x) = \frac{x^2}{x-1}, \quad \text{with } x \neq 1$$

The domain of multiplication is all $\mathbb{R}$, while the domain of division is $\mathbb{R} \setminus \{1\}$, since $g(x) = 0$ when $x = 1$. ■

Lema 7.2.2 If f and g are continuous functions defined on a common set A, then the multiplication $f \cdot g$ is also continuous on A. The division $\frac{f}{g}$ is continuous on A as long as $g(x) \neq 0$ for all $x \in A$.

Demostración. Let f and g be continuous functions on A. The continuity of the multiplication $f \cdot g$ follows from the property of continuity of arithmetic operations on continuous functions. For the division $\frac{f}{g}$, it is continuous as long as $g(x) \neq 0$, since the quotient function is only defined when the denominator is nonzero. Therefore, both operations preserve continuity under the mentioned conditions. ∎

Theorem 7.2.3 The derivative of the product of two functions is obtained using the **product rule**:

$$(f \cdot g)'(x) = f'(x)g(x) + f(x)g'(x)$$

For division, the **quotient rule** applies:

$$\left(\frac{f}{g}\right)'(x) = \frac{f'(x)g(x) - f(x)g'(x)}{g(x)^2}, \quad \text{with } g(x) \neq 0$$

These rules allow calculating the derivatives of functions formed by multiplication or division.

Corollary 7.2.4 If f and g are differentiable functions, then the derivative of their product and division are continuous functions as long as $g(x) \neq 0$ for division. This is because the operations of addition, subtraction, multiplication, and division preserve continuity when the original functions are differentiable.

■ **Example 7.10** Consider the functions $f(x) = x^2$ and $g(x) = e^x$. The **derivative of the product** $(f \cdot g)(x) = x^2 e^x$ is:

$$(f \cdot g)'(x) = f'(x)g(x) + f(x)g'(x) = 2xe^x + x^2 e^x = (2x + x^2)e^x$$

For the **division**, if $h(x) = \frac{x^2}{e^x}$, the derivative is:

$$h'(x) = \frac{2xe^x - x^2 e^x}{e^{2x}} = \frac{x(2-x)}{e^x}$$

This example demonstrates how to apply the product and quotient rules to differentiate more complex functions. ∎

> (R) The multiplication and division of functions allow generating new functions with different properties and behaviors. These operations are fundamental in mathematical analysis, and deriving such functions using the product and quotient rules is an essential tool in calculus.

> **Exercise 7.9** Given the functions $f(x) = \ln(x)$ and $g(x) = x^2$, find the expression for $(f \cdot g)(x)$ and determine the domain of the resulting function.

> **Exercise 7.10** Consider the functions $f(x) = \sin(x)$ and $g(x) = x + 1$. Compute the derivative of the division $\left(\frac{f}{g}\right)(x)$ and determine the domain of the derived function.

> (R) These exercises help readers practice the multiplication and division of functions and apply the corresponding differentiation rules, reinforcing the analysis of domains and continuity properties.

7.2.3 Properties of Operations with Functions

> **Definition 7.2.3** The **operations with functions**, including addition, subtraction, multiplication, and division, possess various fundamental properties similar to those of real numbers. Among these properties are commutativity, associativity, distributivity, and the existence of additive and multiplicative identities. These properties facilitate algebraic operations with functions efficiently and ensure that the results of function combinations retain certain characteristics.

■ **Example 7.11** Consider the functions $f(x) = x + 2$ and $g(x) = x^2 - 1$. The **distributive property** can be observed in the following operation:

$$f(x) \cdot (g(x) + 1) = (x+2) \cdot (x^2 - 1 + 1) = (x+2) \cdot x^2 = x^3 + 2x^2$$

This example shows how the product of a function distributes over the sum of two other functions. ∎

Lema 7.2.3 If f and g are functions defined on a set A, then the addition and multiplication of functions are **commutative**, i.e.:

$$f(x) + g(x) = g(x) + f(x), \quad f(x) \cdot g(x) = g(x) \cdot f(x)$$

for all $x \in A$.

7.2 Operations with Functions (Addition, Subtraction, Multiplication, Division)

Demostración. Let $f(x)$ and $g(x)$ be functions defined on a common set A. By the commutative property of addition and multiplication of real numbers, for all $x \in A$, we have:

$$f(x) + g(x) = g(x) + f(x) \quad \text{and} \quad f(x) \cdot g(x) = g(x) \cdot f(x)$$

Therefore, commutativity holds for the functions f and g on the set A. ∎

> **Theorem 7.2.5** If f, g, and h are functions defined on a common set A, then the **associative property** holds for the addition and multiplication of functions:
>
> $$(f+g)+h = f+(g+h), \quad (f \cdot g) \cdot h = f \cdot (g \cdot h)$$
>
> for all $x \in A$. This property ensures that the grouping order does not affect the result when operating with functions.

Demostración. To demonstrate the associative property of addition and multiplication of functions, consider the sum of functions f, g, and h:

$$[(f+g)+h](x) = (f(x)+g(x))+h(x)$$

By the associative property of real numbers:

$$(f(x)+g(x))+h(x) = f(x)+(g(x)+h(x)) = [f+(g+h)](x)$$

Thus, $(f+g)+h = f+(g+h)$.
For multiplication:

$$[(f \cdot g) \cdot h](x) = (f(x) \cdot g(x)) \cdot h(x)$$

By the associative property of real numbers:

$$(f(x) \cdot g(x)) \cdot h(x) = f(x) \cdot (g(x) \cdot h(x)) = [f \cdot (g \cdot h)](x)$$

Therefore, $(f \cdot g) \cdot h = f \cdot (g \cdot h)$.
This demonstrates that the associative property holds for the addition and multiplication of functions. ∎

> **Corollary 7.2.6** The existence of the **additive identity** and **multiplicative identity** guarantees that, for any function f defined on a set A, the following holds:
>
> $$f(x)+0 = f(x), \quad f(x) \cdot 1 = f(x)$$
>
> where 0 and 1 represent constant functions that take these values for all $x \in A$. These identities are fundamental for operations with functions.

■ **Example 7.12** Let $f(x) = x^2 + 1$. The **additive identity** states:

$$f(x) + 0 = (x^2 + 1) + 0 = x^2 + 1$$

and the **multiplicative identity** states:

$$f(x) \cdot 1 = (x^2 + 1) \cdot 1 = x^2 + 1$$

Thus, the function remains unchanged when adding zero or multiplying by one. ■

> (R) The properties of operations with functions, such as commutativity, associativity, and distributivity, are essential for simplifying algebraic manipulations of functions. Understanding these properties is key to solving complex problems involving multiple functions.

> **Exercise 7.11** Consider the functions $f(x) = 3x - 2$ and $g(x) = x^2 + 1$. Verify the associative property for the sum $(f + g) + h$ with $h(x) = 2x$.

> **Exercise 7.12** Given the functions $f(x) = x^3$ and $g(x) = x + 4$, prove that the distributive property holds for the expression $f(x) \cdot (g(x) + 2)$.

> (R) These exercises allow the reader to apply the properties of operations with functions, verifying results and reinforcing the understanding of the fundamental characteristics governing such operations.

7.3 Piecewise-Defined Functions

7.3.1 Definition of Piecewise Functions

> **Definition 7.3.1** A **piecewise function** is a function defined by different algebraic expressions over distinct intervals of its domain. Formally, a function $f(x)$ is piecewise if:
>
> $$f(x) = \begin{cases} f_1(x), & \text{if } x \in A_1 \\ f_2(x), & \text{if } x \in A_2 \\ \vdots & \vdots \\ f_n(x), & \text{if } x \in A_n \end{cases}$$
>
> where each $f_i(x)$ is a mathematical expression, and each A_i is a subset of the function's domain.

■ **Example 7.13** Consider the piecewise function $f(x)$ defined as follows:

$$f(x) = \begin{cases} x + 2, & \text{if } x < 0 \\ x^2, & \text{if } 0 \leq x \leq 2 \\ 3x - 1, & \text{if } x > 2 \end{cases}$$

This function takes different expressions depending on the value of x, allowing it to model varying behaviors across distinct intervals. ■

7.3 Piecewise-Defined Functions

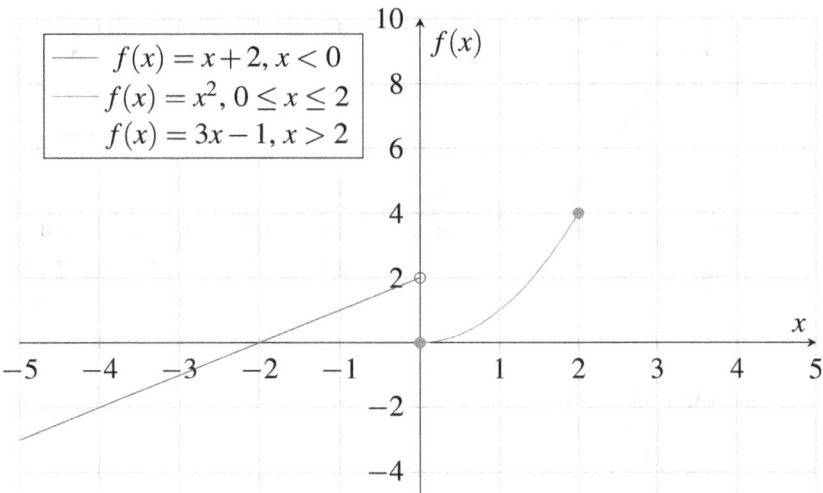

Lema 7.3.1 If each function $f_i(x)$ that defines a piecewise function is continuous on its respective interval A_i, and the endpoints of the intervals are also continuous, then the piecewise function $f(x)$ is continuous on its domain.

Demostración. Let $f(x)$ be a piecewise function defined over distinct intervals $A_1, A_2, \ldots, A_n$, and let each $f_i(x)$ be continuous on A_i. For $f(x)$ to be continuous over its entire domain, it is necessary not only that each $f_i(x)$ is continuous within its interval but also that the limits coincide at the endpoints of the intervals. If this condition holds at every endpoint, then $f(x)$ is continuous throughout its domain. ■

Theorem 7.3.1 A piecewise function can be differentiable if each segment is differentiable within its interval, and the lateral derivatives agree at the endpoints of the intervals. This implies that the differentiability of a piecewise function depends on both the differentiability of each segment and the smoothness at the endpoints.

Corollary 7.3.2 If $f(x)$ is a piecewise function and is differentiable on each of its intervals, and the lateral derivatives agree at the endpoints, then the derivative of $f(x)$ is a continuous function over its entire domain. This result is useful for analyzing the smoothness of segmented functions.

■ **Example 7.14** Let the function $f(x)$ be defined as:

$$f(x) = \begin{cases} 2x+1, & \text{if } x < 1 \\ x^2, & \text{if } x \geq 1 \end{cases}$$

To verify if $f(x)$ is continuous at $x = 1$, compute the limit as x approaches 1 from the left and from the right:

$$\lim_{x \to 1^-} f(x) = 2(1) + 1 = 3, \quad \lim_{x \to 1^+} f(x) = 1^2 = 1$$

Since the limits do not match, the function is not continuous at $x = 1$. This illustrates how discontinuities may arise at the endpoints of a piecewise function. ■

Capítulo 7. Functions: Domain, Range, and Operations

 Piecewise functions are extremely useful for modeling situations where a function exhibits different behaviors depending on the value of the variable. A common example is tax calculation, where the formula varies according to income brackets.

Exercise 7.13 Define a piecewise function that is continuous across its entire domain and consists of two segments, one linear and one quadratic. Verify its continuity at the transition points.

Exercise 7.14 Let $f(x)$ be the piecewise function defined as:

$$f(x) = \begin{cases} x^3, & \text{if } x \leq 0 \\ \sqrt{x}, & \text{if } x > 0 \end{cases}$$

Determine whether $f(x)$ is continuous at $x = 0$.

These exercises help the reader become familiar with the definition of piecewise functions and how to verify fundamental properties such as continuity at transition points.

7.3.2 Graphical Representation of Piecewise-Defined Functions

Definition 7.3.2 A **piecewise-defined function** is a function described by different expressions in distinct intervals of its domain. These functions are graphically represented by plotting each segment separately according to the corresponding expression for each interval. The graphical representation is crucial for visualizing the continuity, differentiability, and overall behavior of the function.

■ **Example 7.15** Consider the piecewise function $f(x)$ defined as follows:

$$f(x) = \begin{cases} -x+2, & \text{if } x < 0 \\ x^2, & \text{if } 0 \leq x < 3 \\ 2x-1, & \text{if } x \geq 3 \end{cases}$$

To graph this function, we plot each segment according to its respective interval. For $x < 0$, the function is a decreasing line with a slope of -1 and y-intercept 2. For $0 \leq x < 3$, the function is a quadratic polynomial, and for $x \geq 3$, the function is a line with a positive slope. This decomposition helps visualize the behavior of $f(x)$ across its domain. ■

7.3 Piecewise-Defined Functions

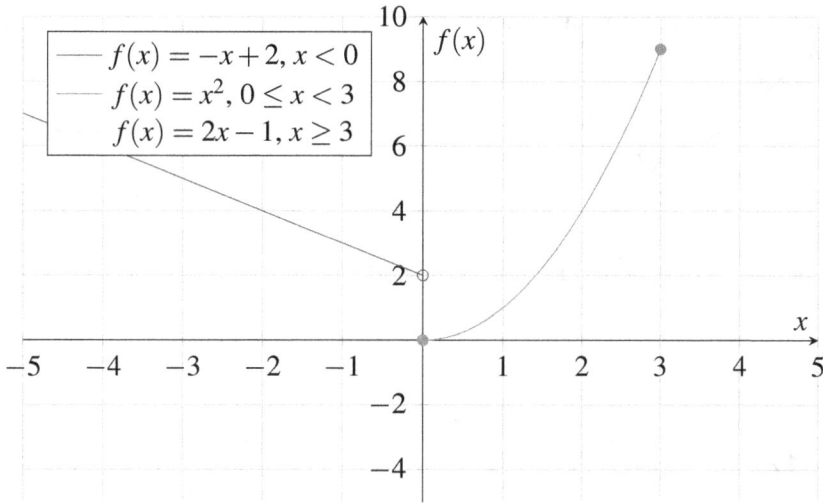

Lema 7.3.2 If each segment of a piecewise-defined function is continuous and the endpoints coincide, then the function is **continuous over its entire domain**.

Demostración. Consider that the function $f(x)$ is defined piecewise over disjoint intervals of its domain and that each segment is continuous in its respective interval. For $f(x)$ to be continuous over its entire domain, the lateral limits must coincide at the endpoints of the intervals. If this condition is satisfied at all endpoints, the overall function is continuous. ∎

Theorem 7.3.3 A **piecewise-defined function** is differentiable over its entire domain if each segment is differentiable and the lateral derivatives agree at the endpoints of the intervals. Furthermore, if the function is continuous but does not satisfy the condition of equal lateral derivatives, it will have a discontinuity in its derivative, even though the function itself remains continuous.

Corollary 7.3.4 If a piecewise-defined function $f(x)$ has segments that are differentiable in their respective intervals and the lateral derivatives agree at the endpoints, then the derivative of $f(x)$ is continuous. This ensures the smoothness of the function and the absence of çorners.°r "kinksïn its graph.

■ **Example 7.16** Let $f(x)$ be defined piecewise as:

$$f(x) = \begin{cases} 2x+3, & \text{if } x < 1 \\ -x^2+4x-3, & \text{if } x \geq 1 \end{cases}$$

To graph this function, plot each segment according to its corresponding expression. For $x < 1$, the function is linear with a slope of 2, and for $x \geq 1$, the function is a quadratic polynomial. At $x = 1$, we must check continuity and differentiability, i.e., whether the lateral limits and derivatives match.

■

> **R** Graphical representation of piecewise-defined functions allows for a visual analysis of continuity and differentiability at critical points. These points often correspond to transitions between different expressions, making it easier to identify potential discontinuities or breaks in the derivative.

Exercise 7.15 Graph the following piecewise function:

$$f(x) = \begin{cases} x^3, & \text{if } x < 0 \\ 2x+1, & \text{if } 0 \leq x < 2 \\ 3-x, & \text{if } x \geq 2 \end{cases}$$

Determine whether $f(x)$ is continuous at $x = 0$ and $x = 2$.

Exercise 7.16 Given the piecewise function $f(x)$:

$$f(x) = \begin{cases} e^x, & \text{if } x < 1 \\ 3x-2, & \text{if } x \geq 1 \end{cases}$$

Verify whether $f(x)$ is differentiable at $x = 1$.

(R) These exercises allow the reader to practice graphing piecewise-defined functions and visually analyze their continuity and differentiability at transition points.

7.3.3 Applications in Real-World Models

Definition 7.3.3 A **mathematical model** is an abstract representation that uses mathematical functions to describe a real-world phenomenon. In particular, **piecewise-defined functions** are used to model situations where behavior changes depending on the interval of values, as is often the case in physics, economics, and biology.

■ **Example 7.17** Consider a model that describes the cost of a phone call with different rates depending on the call duration. Let $C(t)$ represent the cost as a function of time t (in minutes):

$$C(t) = \begin{cases} 0{,}50t, & \text{if } 0 < t \leq 5 \\ 2{,}50 + 0{,}30(t-5), & \text{if } t > 5 \end{cases}$$

This model defines two segments: a cost of 0.50 monetary units per minute for the first 5 minutes, and a fixed cost of 2.50 units plus 0.30 for each additional minute beyond the first 5. This structure is a practical example of how piecewise functions can be used to model tariffs and costs. ■

Lema 7.3.3 If a piecewise function models a physical phenomenon and each segment is continuous, then the mathematical model representing the phenomenon is also continuous. Continuity ensures there are no abrupt changes in values, which is essential for the accuracy of models in many real-world applications.

Theorem 7.3.5 In many practical situations, such as determining tariffs or manufacturing processes, piecewise-defined functions allow for the creation of models that are **piecewise continuous** but not necessarily **piecewise differentiable**. That is, the model can be continuous across its entire domain but may have points where it is not differentiable, such as corners or sharp changes in slope.

7.4 Solved Exercises

Corollary 7.3.6 If a mathematical model is piecewise-defined and its components are linear, the model is easier to manipulate algebraically and analyze graphically. This is common in modeling service tariffs and in linear optimization problems.

> (R) Piecewise representation not only facilitates the modeling of complex situations but also allows for the visualization and analysis of behavior changes that cannot be easily described with a single algebraic expression.

■ **Example 7.18** A classic example of the application of piecewise-defined functions is a progressive tax model. Let $T(I)$ represent the tax rate as a function of income I:

$$T(I) = \begin{cases} 0{,}10I, & \text{if } 0 \leq I \leq 20000 \\ 2000 + 0{,}20(I - 20000), & \text{if } I > 20000 \end{cases}$$

This model describes a tax system where the first 20,000 monetary units are taxed at a rate of 10 ■

Exercise 7.17 Given the piecewise-defined function that models the cost of urban transportation services:

$$C(d) = \begin{cases} 5, & \text{if } 0 < d \leq 10 \text{ km} \\ 5 + 0{,}50(d - 10), & \text{if } d > 10 \text{ km} \end{cases}$$

Determine the cost of a 15 km trip and verify if the cost is continuous at $d = 10$.

Exercise 7.18 Consider an electricity tariff system where the cost depends on monthly energy consumption in kWh. Let $E(C)$ represent the cost as a function of consumption C:

$$E(C) = \begin{cases} 0{,}12C, & \text{if } 0 < C \leq 100 \\ 12 + 0{,}15(C - 100), & \text{if } C > 100 \end{cases}$$

Calculate the cost of consuming 150 kWh and verify if the function is continuous at $C = 100$.

> (R) These exercises allow readers to apply the concept of piecewise functions in real-world problems, such as calculating transportation and electricity tariffs, while verifying continuity at the segment junctions.

7.4 Solved Exercises

Exercise 7.19 Determine the domain and range of the function $f(x) = \frac{2x+3}{x-1}$.

Solution:
- **Domain:** The function $f(x) = \frac{2x+3}{x-1}$ has a restriction in the denominator, as it cannot be equal to zero. Therefore, the domain is:

$$x \neq 1 \Rightarrow D_f = \mathbb{R} \setminus \{1\}$$

- **Range:** There are no restrictions on the resulting values of the function. The function can take any real value, as there is no maximum or minimum value. Hence, the range is:

$$R_f = \mathbb{R}$$

Exercise 7.20 Calculate $(f+g)(x)$ for the functions $f(x) = x^2$ and $g(x) = 3x - 2$.

Solution:
The sum of the functions f and g is given by:

$$(f+g)(x) = f(x) + g(x)$$

Substituting the expressions for $f(x)$ and $g(x)$:

$$(f+g)(x) = x^2 + (3x - 2)$$

$$(f+g)(x) = x^2 + 3x - 2$$

Exercise 7.21 Determine the domain of the function $f(x) = \sqrt{x^2 - 4}$.

Solution:
For the function to be defined, the argument of the square root must be non-negative:

$$x^2 - 4 \geq 0$$

Solving the inequality:

$$(x-2)(x+2) \geq 0$$

The values satisfying the inequality are $x \leq -2$ or $x \geq 2$. Therefore, the domain is:

$$D_f = (-\infty, -2] \cup [2, \infty)$$

Exercise 7.22 Calculate $(f \cdot g)(x)$ for $f(x) = 2x$ and $g(x) = x + 1$.

Solution:
The product of the functions f and g is given by:

$$(f \cdot g)(x) = f(x) \cdot g(x)$$

Substituting the expressions for $f(x)$ and $g(x)$:

$$(f \cdot g)(x) = (2x)(x+1)$$

$$(f \cdot g)(x) = 2x^2 + 2x$$

Exercise 7.23 Graph the piecewise-defined function:

$$f(x) = \begin{cases} x+2 & \text{if } x < 0 \\ x^2 & \text{if } x \geq 0 \end{cases}$$

Solution:
- For $x < 0$, the function is defined as $f(x) = x+2$, which is a straight line with slope 1 and y-intercept 2. - For $x \geq 0$, the function is defined as $f(x) = x^2$, which is a parabola opening upwards with its vertex at the origin.
To graph:
- When $x < 0$, the graph is a line crossing the y-axis at $y = 2$ with a positive slope. - When $x \geq 0$, the graph is a parabola with its vertex at $(0,0)$.

7.5 Proposed Exercises

7.5.1 Domain and Range of Real Functions

Exercise 7.24 Determine the domain of the function $f(x) = \frac{1}{x-2}$.

Exercise 7.25 Find the range of the function $f(x) = \sqrt{x+4}$.

Exercise 7.26 Determine the domain and range of the function $f(x) = \frac{\sqrt{x-1}}{x+3}$.

Exercise 7.27 Calculate the domain of the function $f(x) = \ln(x^2 - 4)$.

Exercise 7.28 Find the range of the function $f(x) = 3x^2 + 2x - 1$.

7.5.2 Operations with Functions (Sum, Subtraction, Multiplication, Division)

Exercise 7.29 Calculate $(f+g)(x)$ for $f(x) = 2x+3$ and $g(x) = x^2 - 1$.

Exercise 7.30 Find $(f-g)(x)$ for $f(x) = 4x$ and $g(x) = x^2 + 2$.

Exercise 7.31 Determine $(fg)(x)$ for $f(x) = x+1$ and $g(x) = 3x - 4$.

Exercise 7.32 Calculate $\left(\frac{f}{g}\right)(x)$ for $f(x) = x^2 - 1$ and $g(x) = x+2$.

Exercise 7.33 Find the value of $(fg)(2)$ for $f(x) = x+3$ and $g(x) = x-1$.

7.5.3 Piecewise-Defined Functions

Exercise 7.34 Define $f(x)$ as follows: $f(x) = \begin{cases} x^2 & \text{if } x < 0 \\ 2x+1 & \text{if } x \geq 0 \end{cases}$. Find $f(-2)$ and $f(3)$.

Exercise 7.35 Determine the domain of the piecewise-defined function: $f(x) = \begin{cases} \sqrt{x} & \text{if } x \geq 0 \\ -x & \text{if } x < 0 \end{cases}$.

Exercise 7.36 Graph the function defined by $f(x) = \begin{cases} x+2 & \text{if } x < 1 \\ x^2-1 & \text{if } x \geq 1 \end{cases}$.

Exercise 7.37 Find the value of $f(x)$ when $x = 1$ for the piecewise-defined function $f(x) = \begin{cases} 3x & \text{if } x < 2 \\ x^2 & \text{if } x \geq 2 \end{cases}$.

Exercise 7.38 Determine whether the function $f(x) = \begin{cases} 2x & \text{if } x < 0 \\ 3x+1 & \text{if } x \geq 0 \end{cases}$ is continuous at $x = 0$.

8. Graphs and Graphical Transformations

8.1 Translation and Reflection of Functions

8.1.1 Vertical and Horizontal Translation

Definition 8.1.1 The **translation of functions** refers to the shifting of a function's graph along the coordinate axes. There are two main types of translations: **vertical** and **horizontal**. A vertical translation involves moving the graph up or down, while a horizontal translation involves moving it to the right or left.

■ **Example 8.1** Consider the function $f(x) = x^2$. If we apply a vertical translation upwards by 3 units, the new function is $g(x) = x^2 + 3$. If we apply a horizontal translation 2 units to the right, the new function is $h(x) = (x-2)^2$. These translations affect the position of the graph without altering its shape as a parabola. ■

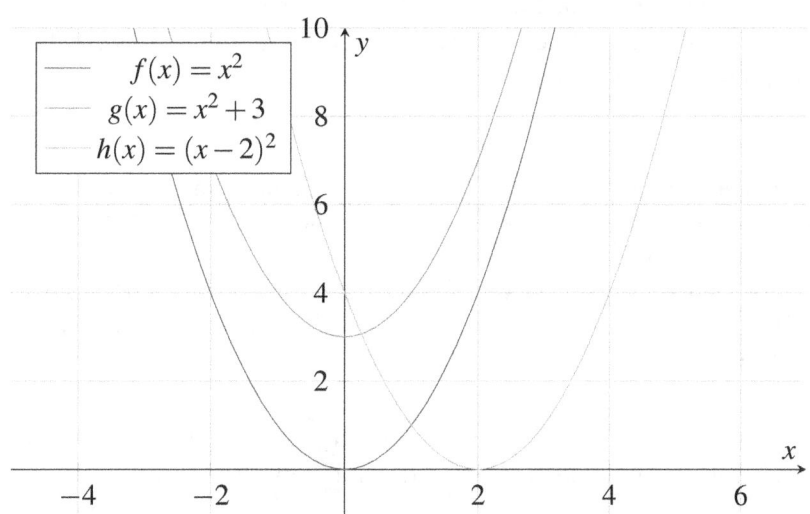

Lema 8.1.1 If a function $f(x)$ is continuous, then a translation of $f(x)$ (either vertical or horizontal) will also be continuous. This means that the continuity properties of a function are preserved under

translation.

Demostración. Let $f(x)$ be a continuous function. Consider two types of translations:
1. **Horizontal Translation**: $g(x) = f(x-h)$, where $h \in \mathbb{R}$ is a constant.
2. **Vertical Translation**: $g(x) = f(x) + k$, where $k \in \mathbb{R}$ is a constant.

To show that $g(x)$ is continuous in each case:

- For a **horizontal translation**: Since $f(x)$ is continuous, for any point $x_0 \in \mathbb{R}$, the limit $\lim_{x \to x_0} f(x-h) = f(x_0 - h)$. As f is continuous, $\lim_{x \to x_0} f(x-h) = g(x_0)$, showing that $g(x)$ is also continuous.
- For a **vertical translation**: The vertical translation simply adds a constant k to the value of $f(x)$. Since $f(x)$ is continuous, $\lim_{x \to x_0} [f(x) + k] = \lim_{x \to x_0} f(x) + k = f(x_0) + k$. This shows that $g(x)$ is continuous.

Therefore, in both cases, the resulting function $g(x)$ is continuous, proving that a translation of a continuous function is also continuous. ∎

> **Theorem 8.1.1** Let $f(x)$ be a function defined on a set $D \subseteq \mathbb{R}$. If we perform a horizontal translation of c units to the right, the new function $g(x) = f(x-c)$ has the same domain but with its values shifted. Similarly, a vertical translation of k units upwards is expressed as $h(x) = f(x) + k$, which modifies the range but retains the same domain.

Demostración. Let $f(x)$ be a function defined on a domain $D \subseteq \mathbb{R}$.
1. **Horizontal Translation** ($g(x) = f(x-c)$): - The function $g(x) = f(x-c)$ is obtained by shifting each input value of $f(x)$ by c units to the right. - Since $f(x)$ has domain D, the function $g(x)$ will have the same domain. That is, if $x \in D$, then $x - c \in D$ implies $x \in D$ for $g(x)$. - Hence, the domain of $g(x)$ is the same as that of $f(x)$.
2. **Vertical Translation** ($h(x) = f(x) + k$): - The function $h(x) = f(x) + k$ is obtained by adding k units to the output of $f(x)$. This operation does not affect the domain, as it does not depend on the value of x. - However, the range of $h(x)$ will differ from that of $f(x)$, as each value of $f(x)$ is increased by k units. - The domain of $h(x)$ remains the same as that of $f(x)$, but the range changes to $R_h = R_f + k$, where R_f is the range of $f(x)$.

In summary, horizontal and vertical translations retain the original domain, while vertical translations alter the range of the function. This completes the proof of the theorem. ∎

> **Corollary 8.1.2** Translating a function does not alter the nature of its growth or decay. In other words, if $f(x)$ is an increasing function, then any translation of $f(x)$ will also be increasing. Similarly, this applies to decreasing functions.

> (R) Translations are useful in constructing and transforming mathematical models, allowing the adjustment of function graphs to different contexts without changing their essential behavior.

■ **Example 8.2** Consider the linear function $f(x) = 2x + 1$. Applying a horizontal translation of 3 units to the left gives $g(x) = 2(x+3) + 1 = 2x + 7$. Similarly, a vertical translation downwards by 2 units results in $h(x) = 2x + 1 - 2 = 2x - 1$. ■

> **Exercise 8.1** Given the function $f(x) = \sqrt{x}$, find the expression for the new function after applying a translation of 4 units to the right and 2 units upwards.

8.1 Translation and Reflection of Functions

Exercise 8.2 Let $f(x) = \ln(x)$, apply a translation of 3 units to the left and determine the expression for the new function. Additionally, discuss how the domain of the function changes after the translation.

(R) These exercises help to understand how translations affect the overall behavior of functions and enable visualization of the changes in their graphical representations.

8.1.2 Reflection with Respect to the Axes

Definition 8.1.2 The **reflection of a function with respect to an axis** is a geometric transformation that produces a mirror image of the function's graph. There are two main types of reflections: **reflection with respect to the x-axis** and **reflection with respect to the y-axis**. Reflection with respect to the x-axis is achieved by changing the sign of the function's values, while reflection with respect to the y-axis is achieved by changing the sign of the function's argument.

■ **Example 8.3** Consider the function $f(x) = x^2$. The reflection with respect to the x-axis is given by $g(x) = -f(x) = -x^2$. The reflection with respect to the y-axis is given by $h(x) = f(-x) = (-x)^2 = x^2$, which coincides with the original function due to the symmetry of the parabola with respect to the y-axis. This example illustrates how reflections can modify or maintain a function's graph depending on its symmetry. ■

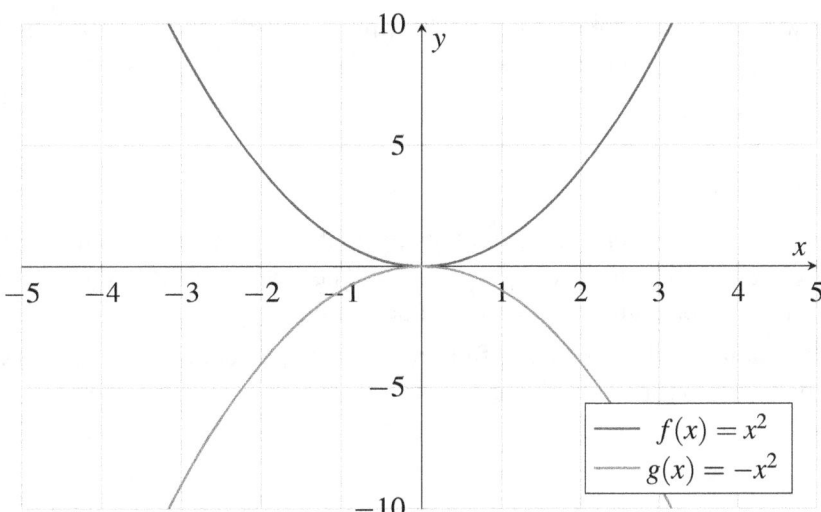

Lema 8.1.2 If $f(x)$ is an even function, then reflection with respect to the y-axis does not alter its graph. This is because a function is even if $f(x) = f(-x)$ for all x in its domain, meaning its graph is already symmetric with respect to the y-axis.

Demostración. By definition, a function $f(x)$ is **even** if it satisfies the condition:

$$f(x) = f(-x) \quad \text{for all } x \text{ in its domain.}$$

This implies that the value of the function at a point x is equal to its value at the opposite point $-x$. In other words, the graph of $f(x)$ is symmetric with respect to the y-axis because for every point $(x, f(x))$, there exists another point $(-x, f(-x))$ with the same function value.

If we reflect the graph of $f(x)$ with respect to the y-axis, each point $(x, f(x))$ is transformed into $(-x, f(x))$, which already belongs to the graph due to the evenness condition. Therefore, the graph of the function remains unchanged under this reflection.

This proves that if $f(x)$ is an even function, reflection with respect to the y-axis does not change its graph. ∎

> **Theorem 8.1.3** Let $f(x)$ be a function defined on $\mathbb{R}$. If $f(x)$ is reflected with respect to the x-axis, the new function is $g(x) = -f(x)$. If $f(x)$ is reflected with respect to the y-axis, the new function is $h(x) = f(-x)$. These transformations preserve the domain of the function but may alter the range depending on the behavior of $f(x)$.

Demostración. To prove this theorem, consider the two reflection transformations:

1. **Reflection with respect to the x-axis**: Reflecting the function $f(x)$ with respect to the x-axis results in the new function $g(x)$ defined as:

$$g(x) = -f(x).$$

This transformation changes the sign of each point on the graph of $f(x)$, reflecting it across the x-axis. The domain remains unchanged because the values of x are not affected. However, the range changes as positive values become negative and vice versa.

2. **Reflection with respect to the y-axis**: Reflecting the function $f(x)$ with respect to the y-axis results in the new function $h(x)$ defined as:

$$h(x) = f(-x).$$

This transformation replaces each x-value with its opposite $-x$. Again, the domain of $f(x)$ remains unchanged since we are only substituting x with $-x$. The graph is reflected horizontally, altering the function's behavior depending on its original definition.

In both cases, the domain of the original function $f(x)$ is preserved, but the range may change depending on the type of reflection. ∎

> **Corollary 8.1.4** If a function $f(x)$ is odd, then reflecting it with respect to the x-axis or the y-axis results in an equivalent transformation. That is, $-f(-x) = f(x)$ for an odd function. This implies that odd functions exhibit rotational symmetry about the origin.

> (R) Reflections of functions with respect to the axes are important transformations for analyzing the symmetry and geometric properties of functions. They help identify special characteristics that simplify mathematical problem-solving.

■ **Example 8.4** Consider the linear function $f(x) = 2x + 3$. The reflection with respect to the x-axis is $g(x) = -f(x) = -2x - 3$, which reverses the slope and the intercept. The reflection with respect to the y-axis is $h(x) = f(-x) = -2x + 3$, which reverses only the slope. ■

8.1 Translation and Reflection of Functions

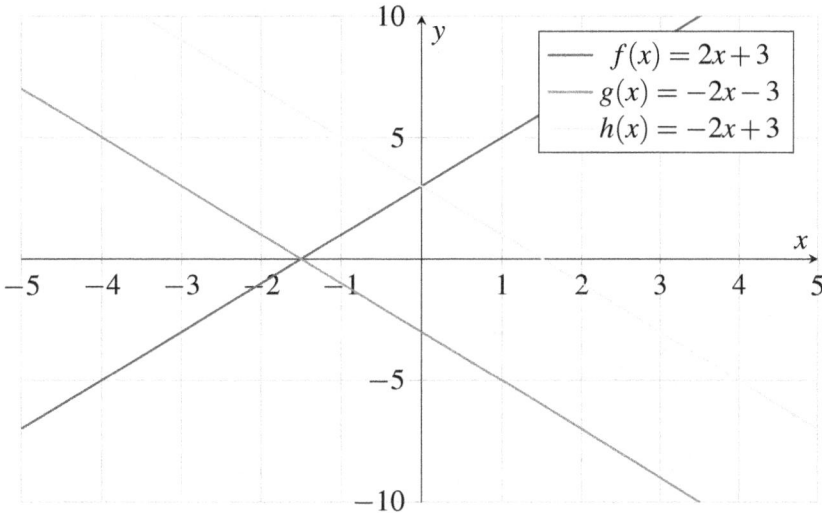

Exercise 8.3 Given the function $f(x) = \sin(x)$, find the expression of the new function after a reflection with respect to the y-axis and verify whether the new function is even, odd, or neither.

Exercise 8.4 Consider the function $f(x) = e^x$. Perform a reflection with respect to the x-axis and determine how the range of the function changes.

(R) These exercises help to understand how reflections with respect to the axes affect the overall behavior of functions and analyze the resulting symmetry.

8.1.3 Graphical Representation of Transformations

Definition 8.1.3 A **graphical transformation** of a function involves modifying the shape or position of the original function's graph through a series of operations such as translations, reflections, stretches, and compressions. Each transformation has a predictable effect on the graph, either shifting, mirroring, or resizing it.

■ **Example 8.5** Consider the original function $f(x) = x^2$, which has a parabolic shape. Applying the following transformations:

1. Vertical translation upward by 3 units: $g(x) = x^2 + 3$. 2. Reflection with respect to the x-axis: $h(x) = -x^2$. 3. Horizontal compression by a factor of 2: $k(x) = (2x)^2$.

Each of these transformations has a specific visual impact on the parabola's graph, altering its position or shape without changing its fundamental structure. ■

Capítulo 8. Graphs and Graphical Transformations

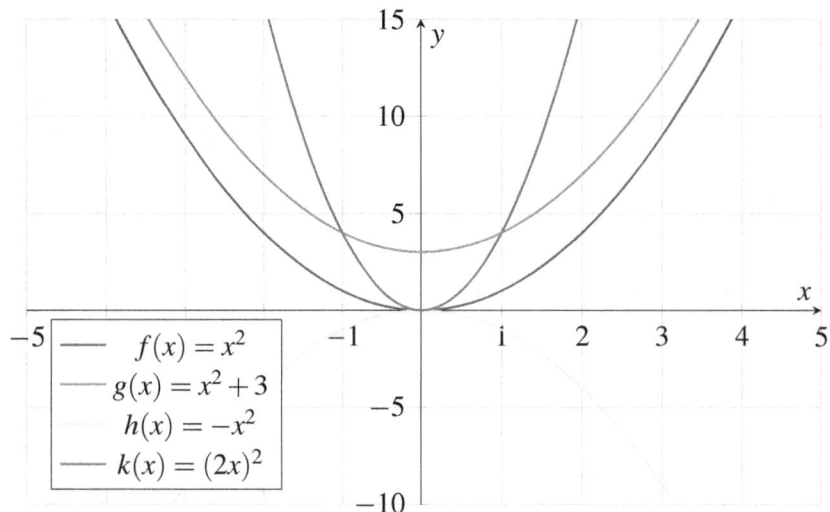

Lema 8.1.3 The composition of two or more transformations on a function $f(x)$ can be graphically represented by sequentially applying each transformation. In particular, the order in which translations and stretches are applied can influence the final result, which is crucial for correctly interpreting the graph.

Theorem 8.1.5 Let $f(x)$ be a function defined over a domain D. If a transformation consisting of a horizontal translation by c units followed by a reflection with respect to the y-axis is applied, the new function can be expressed as $g(x) = f(-x-c)$. This expression illustrates how different types of transformations combine to affect the graphical representation of a function.

Corollary 8.1.6 If a function $f(x)$ is continuous and is subjected to a transformation composed of reflections and translations, the transformed function is also continuous. Continuity is preserved because these operations do not introduce points of discontinuity.

Demostración. Let $f(x)$ be a continuous function, meaning that for all $c \in \mathbb{R}$, the following holds:

$$\lim_{x \to c} f(x) = f(c).$$

Consider a transformation of $f(x)$ that includes reflections and translations. The basic operations to analyze are:
1. **Reflection with respect to the y-axis**: $g(x) = f(-x)$. - Since f is continuous at c, then:

$$\lim_{x \to c} f(-x) = f(-c) = g(c).$$

This implies that $g(x)$ is continuous at c.
2. **Reflection with respect to the x-axis**: $h(x) = -f(x)$. - Since f is continuous at c, then:

$$\lim_{x \to c} -f(x) = -\lim_{x \to c} f(x) = -f(c) = h(c).$$

This implies that $h(x)$ is continuous at c.
3. **Translation along the x-axis**: $k(x) = f(x-a)$, with $a \in \mathbb{R}$. - Since f is continuous at $c+a$, then:

$$\lim_{x \to c} f(x-a) = f(c) = k(c).$$

8.2 Vertical and Horizontal Scaling

This implies that $k(x)$ is continuous at c.

4. **Translation along the y-axis**: $m(x) = f(x) + b$, with $b \in \mathbb{R}$. - Since f is continuous at c, then:

$$\lim_{x \to c}(f(x) + b) = \lim_{x \to c} f(x) + b = f(c) + b = m(c).$$

This implies that $m(x)$ is continuous at c.

Since each of these transformations preserves continuity, any composition of them will also preserve continuity. Therefore, the transformed function is continuous. ∎

> (R) Understanding graphical transformations and their representation is essential for visualizing and comprehending how functions behave in different contexts, enabling mathematical models to be adapted to practical situations.

■ **Example 8.6** Consider the function $f(x) = \cos(x)$. Applying a horizontal translation to the right by $\frac{\pi}{2}$ and a reflection with respect to the x-axis results in the transformed function $g(x) = -\cos(x - \frac{\pi}{2}) = \sin(x)$. This example demonstrates how a combination of transformations can convert one trigonometric function into another. ■

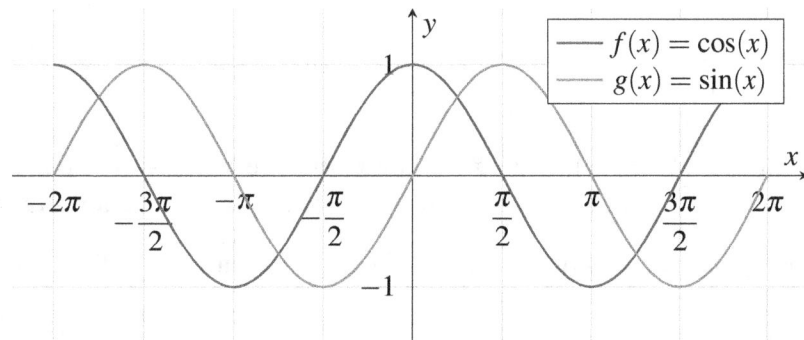

> Exercise 8.5 Given the function $f(x) = e^x$, perform a vertical translation downward by -3 units and a reflection with respect to the y-axis. Write the expression of the new function and discuss how its domain and range change.

> Exercise 8.6 Let $f(x) = |x|$. Apply a horizontal compression by a factor of 2 and then a vertical translation upward by 4 units. Find the expression of the new transformed function and describe its main characteristics.

> (R) These exercises allow the reader to apply the concepts of graphical transformations and observe how each type of transformation affects the domain, range, and visual representation of a function.

8.2 Vertical and Horizontal Scaling

8.2.1 Scaling Functions Upward and Downward

> **Definition 8.2.1 Scaling functions** is a transformation where all values of a function are multiplied by a constant. If the function $f(x)$ is multiplied by a positive factor $k > 1$, we obtain

an **upward scaling**, meaning the graph of the function is "stretched" vertically. If $0 < k < 1$, a **downward scaling** occurs, where the graph is çompressed" vertically.

■ **Example 8.7** Consider the function $f(x) = x^2$. Applying an upward scaling by multiplying by $k = 2$ results in $g(x) = 2x^2$, which "stretches" the graph vertically. Applying a downward scaling by multiplying by $k = \frac{1}{2}$ results in $h(x) = \frac{1}{2}x^2$, which çompresses" the graph vertically. These changes affect the parabola's opening, making it "narrower." or "wider" depending on the value of k. ■

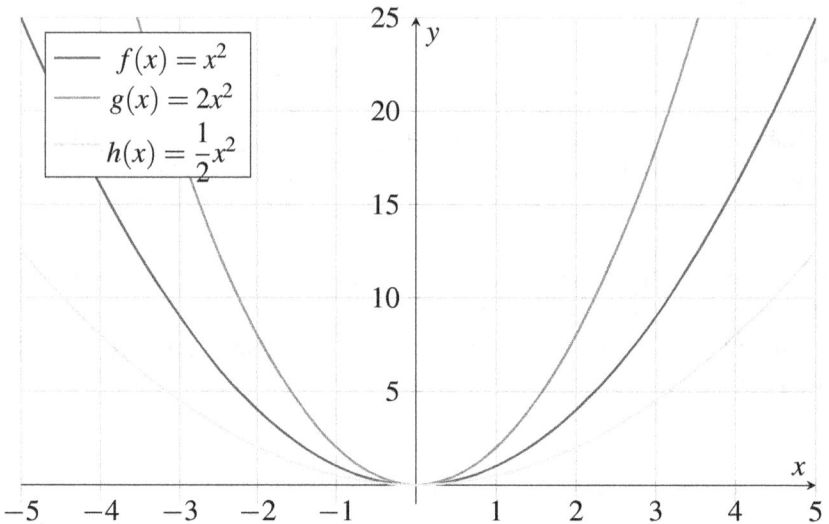

Lema 8.2.1 For any function $f(x)$, scaling by multiplying the function by a constant $k > 0$ does not change the domain of $f(x)$. Scaling only affects the values of the function's range.

Demostración. Let $f(x)$ be a function with domain $D \subseteq \mathbb{R}$. Consider scaling $f(x)$ by a positive constant $k > 0$:

$$g(x) = k \cdot f(x).$$

The domain of $f(x)$ is the set of values of x for which $f(x)$ is defined, i.e., $x \in D$.
The function $g(x) = k \cdot f(x)$ is defined whenever $f(x)$ is defined because multiplying by the constant k introduces no additional restrictions on x. Therefore, the domain of $g(x)$ is the same as the domain of $f(x)$:

$$\text{dom}(g) = \text{dom}(f).$$

Scaling only affects the function's image, multiplying each value of $f(x)$ by k, which changes the range of the function but not its domain.

■

Theorem 8.2.1 Let $f(x)$ be a function defined on a domain D. If an upward scaling by a constant $k > 1$ is applied, the range of the new function $g(x) = kf(x)$ is expanded by a factor of k. If $0 < k < 1$, the range is compressed by the same factor.

Demostración. Let $f(x)$ be a function with domain D and range R, such that:

$$R = \{f(x) \mid x \in D\}.$$

Consider the new function $g(x) = kf(x)$, where $k > 0$.

8.2 Vertical and Horizontal Scaling

1. **Case 1: $k > 1$**
- If $y \in R$, then there exists $x \in D$ such that $f(x) = y$. - For the scaled function, $g(x) = kf(x) = ky$. - Since $k > 1$, the value of $g(x)$ is expanded by the factor k. Thus, the new range R_g of $g(x)$ is:

$$R_g = \{ky \mid y \in R\}.$$

This means all values of the range of $f(x)$ are multiplied by k, expanding the range by a factor of k.
2. **Case 2: $0 < k < 1$**
- If $y \in R$, then there exists $x \in D$ such that $f(x) = y$. - For the scaled function, $g(x) = kf(x) = ky$. - Since $0 < k < 1$, the value of $g(x)$ is compressed by the factor k. Thus, the new range R_g of $g(x)$ is:

$$R_g = \{ky \mid y \in R\}.$$

This means all values of the range of $f(x)$ are multiplied by k, compressing the range by a factor of k.

In both cases, multiplying by k only affects the image of the function, either expanding or compressing the range depending on k.

∎

Corollary 8.2.2 If $f(x)$ is a function bounded above and an upward scaling is applied, the scaled function remains bounded, but its upper bound increases by a factor of k. Similarly, a downward scaling decreases the upper bound. This behavior is significant in analyzing bounded functions and their applications.

Demostración. Suppose $f(x)$ is a function bounded above, meaning there exists a real number $M > 0$ such that:

$$f(x) \leq M \quad \text{for all } x \in D.$$

Now, consider the scaled function $g(x) = kf(x)$ with $k > 0$.
1. **Case 1: $k > 1$ (upward scaling)**
- Multiplying the inequality $f(x) \leq M$ by k, we get:

$$kf(x) \leq kM \quad \text{for all } x \in D.$$

- This implies that the new function $g(x)$ is bounded above by kM.
Therefore, an upward scaling increases the upper bound of the function by a factor of k.
2. **Case 2: $0 < k < 1$ (downward scaling)**
- Multiplying the inequality $f(x) \leq M$ by k, where $0 < k < 1$, we get:

$$kf(x) \leq kM \quad \text{for all } x \in D.$$

- This implies that the new function $g(x)$ is bounded above by kM, where $kM < M$.
Therefore, a downward scaling decreases the upper bound of the function by a factor of k.
In both cases, the scaled function remains bounded, and its new upper bound depends on the scaling constant k.

∎

> (R) Scaling functions has practical applications in various fields, such as physics and economics, where changes in the magnitude of a phenomenon need to be modeled without altering its overall behavior.

■ **Example 8.8** Consider the trigonometric function $f(x) = \sin(x)$. Applying an upward scaling of 3 results in the transformed function $g(x) = 3\sin(x)$. This scaling increases the function's amplitude from 1 to 3, maintaining its frequency and periodicity unchanged. ■

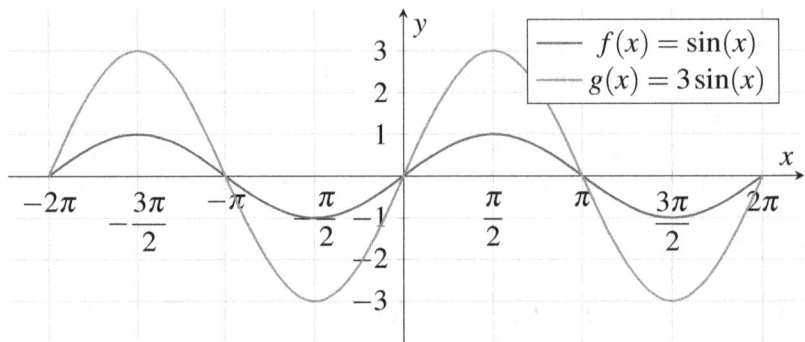

Exercise 8.7 Given the function $f(x) = |x|$, perform a downward scaling by multiplying it by $k = \frac{1}{3}$. Write the expression for the new function and describe how its general behavior changes.

Exercise 8.8 Let $f(x) = e^x$. Perform an upward scaling by 4 units and discuss how the function's range changes. Does the domain change due to this scaling?

(R) These exercises allow the reader to apply the concept of scaling to observe how amplitude and range of a function change, facilitating the understanding of vertical transformations in function analysis.

8.2.2 Horizontal Scaling and Its Effects

Definition 8.2.2 Horizontal scaling is a transformation that modifies the width of a function's graph. It involves multiplying the variable x by a factor k, which affects the frequency and amplitude of the function along the horizontal axis. If $0 < k < 1$, the result is an **outward horizontal scaling**, causing the function to expand along the x-axis. If $k > 1$, the result is an **inward horizontal scaling**, compressing the function.

■ **Example 8.9** Consider the function $f(x) = \sin(x)$. Applying a horizontal scaling by a factor of $k = 2$ results in the transformed function $g(x) = \sin(\frac{x}{2})$. This doubles the original function's period, ."expanding"the graph along the x-axis. Conversely, if $k = \frac{1}{2}$, the resulting function is $h(x) = \sin(2x)$, and the period is halved, çompressing"the graph along the x-axis. ■

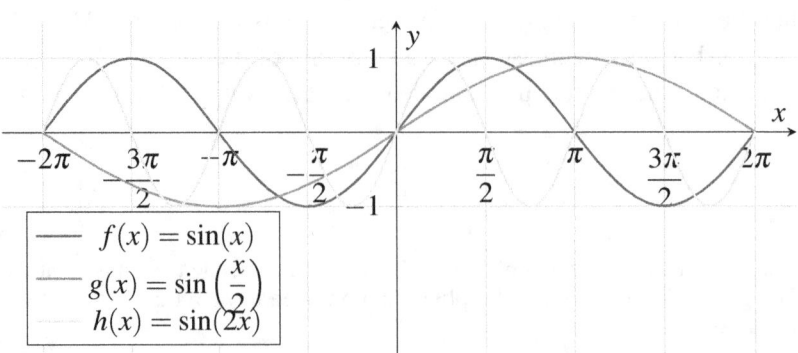

8.2 Vertical and Horizontal Scaling

Lema 8.2.2 For any function $f(x)$, a horizontal scaling by a factor k affects the function's frequency. If $0 < k < 1$, the function has a lower frequency, while if $k > 1$, the frequency increases. This property is particularly important in the analysis of periodic functions.

Demostración. Let $f(x)$ be a periodic function with period T, such that:

$$f(x+T) = f(x) \quad \text{for all } x.$$

Consider a new function $g(x) = f(kx)$, where $k > 0$.
1. **Period of the scaled function**: - To find the period of $g(x)$, determine a value T' such that:

$$g(x+T') = g(x).$$

- Substituting into $g(x)$:

$$g(x+T') = f(k(x+T')) = f(kx + kT').$$

- For $f(kx + kT') = f(kx)$, it is required that $kT' = T$. Thus:

$$T' = \frac{T}{k}.$$

2. **Frequency analysis**: - The frequency of a periodic function is the reciprocal of its period, i.e., Frequency $= \frac{1}{T}$. - For $g(x)$, the new frequency is:

$$\text{Frequency of } g(x) = \frac{1}{T'} = \frac{k}{T}.$$

- If $0 < k < 1$, then $T' > T$, meaning $g(x)$ has a longer period and thus a lower frequency. - If $k > 1$, then $T' < T$, meaning $g(x)$ has a shorter period and thus a higher frequency.
Therefore, a horizontal scaling by a factor k affects a periodic function's frequency, decreasing it if $0 < k < 1$ and increasing it if $k > 1$.
∎

Theorem 8.2.3 Let $f(x)$ be a function defined on a domain D. If a horizontal scaling is performed by multiplying the variable by a factor $k > 0$, the new function takes the form $g(x) = f(kx)$. In this case, if $k > 1$, the domain of the new function shrinks; if $0 < k < 1$, the domain expands proportionally.

Demostración. Let $f(x)$ be a function defined on a domain $D \subseteq \mathbb{R}$, and consider the horizontally scaled function $g(x) = f(kx)$, where $k > 0$.
To analyze how horizontal scaling affects $g(x)$'s domain, consider the set of x-values for which $g(x)$ is defined:
1. **Domain of $g(x)$**: - $f(kx)$ is defined if $kx \in D$. - This implies x must satisfy:

$$x \in \left\{ \frac{d}{k} \mid d \in D \right\}.$$

- If D is the interval $[a,b]$, then the domain of $g(x)$ becomes:

$$\left[\frac{a}{k}, \frac{b}{k} \right].$$

2. **Case 1: $k > 1$**: - If $k > 1$, then $\frac{a}{k} > a$ and $\frac{b}{k} < b$. - This implies the new interval $\left[\frac{a}{k}, \frac{b}{k}\right]$ is narrower than the original interval $[a,b]$. - Thus, the domain of $g(x)$ shrinks when $k > 1$.

3. **Case 2: $0 < k < 1$**: - If $0 < k < 1$, then $\frac{a}{k} < a$ and $\frac{b}{k} > b$. - This implies the new interval $\left[\frac{a}{k}, \frac{b}{k}\right]$ is wider than the original interval $[a,b]$. - Thus, the domain of $g(x)$ expands when $0 < k < 1$. In conclusion, a horizontal scaling of the form $g(x) = f(kx)$ modifies the original function's domain: if $k > 1$, the domain shrinks; if $0 < k < 1$, the domain expands proportionally. ∎

Corollary 8.2.4 If $f(x)$ is a continuous function on its domain, the horizontally scaled function $g(x) = f(kx)$ will also be continuous on its transformed domain. This means the function's continuity is preserved under horizontal scaling, even though its geometric behavior may change.

R Horizontal scaling is a fundamental tool in analyzing periodic functions, especially in areas such as wave physics and signal analysis, where frequency manipulation is crucial for interpreting and modifying function behavior.

■ **Example 8.10** Consider the quadratic function $f(x) = x^2$. Applying a horizontal scaling factor of $k = 3$, the new function becomes $g(x) = (x/3)^2 = \frac{1}{9}x^2$. This transformation expands the parabola along the horizontal axis, "flattening" the original shape. ■

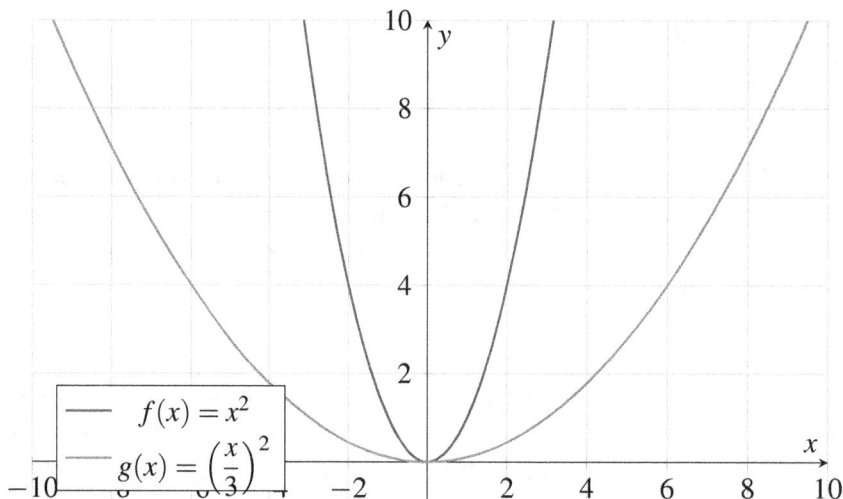

Exercise 8.9 Given the function $f(x) = e^x$, apply an inward horizontal scaling with a factor of $k = 2$. Write the expression for the new function and analyze how the graph's behavior changes compared to the original function.

Exercise 8.10 Let $f(x) = \cos(x)$. Apply a horizontal scaling with a factor of $k = 0,5$ and determine the new function. Describe how the scaling affects the function's periodicity.

R These exercises help understand how horizontal scaling alters a function's behavior, affecting its amplitude and periodicity both visually and analytically.

8.2.3 Applications of Scaling in Graphs

8.2 Vertical and Horizontal Scaling

Definition 8.2.3 Scaling in graphs is a mathematical transformation that involves modifying the size of a function along the vertical or horizontal axis. This transformation can be used to enlarge or reduce the graphical representation of a function without altering its basic shape, affecting only its size or amplitude. Scaling is particularly useful when adjusting the magnitude of a function to analyze trends, patterns, or models in different contexts.

■ **Example 8.11** Consider the function $f(x) = x^3$. If we apply a vertical scaling upward by multiplying by $k = 3$, the transformed function becomes $g(x) = 3x^3$. In this case, the graph of the function is "stretched."along the vertical axis, affecting the amplitude of each point on the graph but not the domain of the function. This technique is especially useful to emphasize the influence of certain variables in a mathematical model or to adjust the level of detail in a graph. ■

Lema 8.2.3 For any function $f(x)$, if scaling is performed along the horizontal or vertical axis, the continuity of the function remains unaffected. However, the values of the range or domain will change depending on the scaling factor used.

Theorem 8.2.5 Let $f(x)$ be a function defined on a domain D. If vertical scaling is applied by multiplying by a factor $k > 0$, the scaled function $g(x) = kf(x)$ will have a range R_g such that $R_g = k \cdot R_f$, where R_f is the range of the original function. Similarly, if scaling is performed on the horizontal axis with a factor $k > 0$, the domain of the function will be proportionally affected.

Corollary 8.2.6 The effect of scaling on a periodic function does not change its periodic nature; however, the amplitude and period of the function may be modified. In the case of trigonometric functions, this is particularly evident in the transformation of the amplitude or frequency of waves.

(R) Scaling is widely used in sciences to adjust mathematical models to appropriate measurement units, making it easier to compare different phenomena. It also has practical applications in engineering and data visualization, where adjusting a graph's scale is necessary to highlight specific aspects of the data.

■ **Example 8.12** Consider the function $f(x) = \sin(x)$. Applying horizontal scaling by a factor of 2, we obtain $g(x) = \sin\left(\frac{x}{2}\right)$. In this case, the function's period doubles, so the graph "stretches."along the horizontal axis, which can be useful for visualizing low-frequency oscillations in signal analysis. ■

Exercise 8.11 Let $f(x) = \cos(x)$. Apply a vertical downward scaling by a factor of $k = \frac{1}{2}$ and write the expression of the new function. How does the range of the function change? ■

Exercise 8.12 Given the function $f(x) = x^2$, perform horizontal scaling inward with a factor of $k = 3$. Explain how the graph's shape is modified and how it affects the domain. ■

(R) These exercises allow exploration of how scaling affects the graphical representation of a function and provide insights into how mathematical transformations relate to their graphical applications. This understanding is fundamental for analysis and modeling in various scientific disciplines.

8.3 Graphs of Composite Functions

8.3.1 Definition of Function Composition

Definition 8.3.1 **Function composition** is an operation that takes two functions f and g and creates a new function h such that $h(x) = f(g(x))$. The function g is applied first to the input value, and then the function f is applied to the result of $g(x)$. The composition of f and g is denoted as $f \circ g$.

■ **Example 8.13** Let $f(x) = x^2 + 1$ and $g(x) = 2x + 3$. The composition $(f \circ g)(x)$ is obtained by evaluating $g(x)$ first and then applying f to the result:

$$(f \circ g)(x) = f(g(x)) = f(2x+3) = (2x+3)^2 + 1$$

This results in a new function. ■

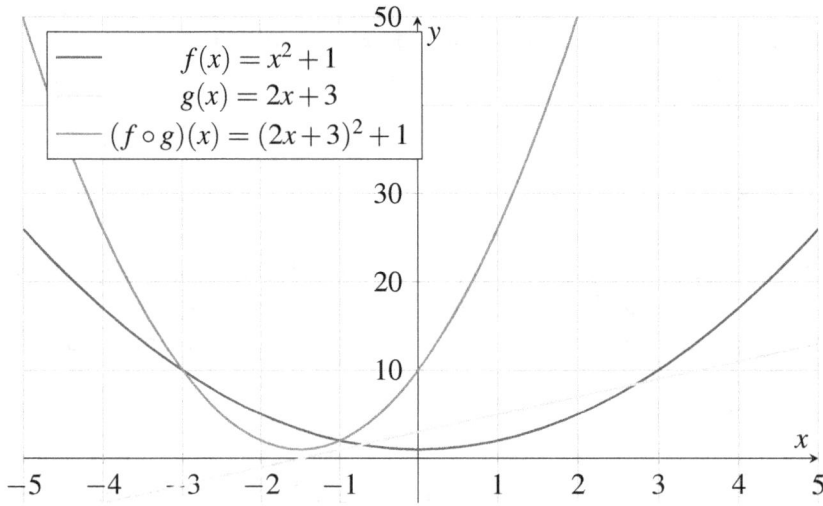

Lema 8.3.1 The composition of functions is not commutative, i.e., in general $f \circ g \neq g \circ f$. This implies that the order in which the functions are applied matters and affects the result of the composition.

Demostración. To demonstrate that function composition is not commutative, consider the following counterexample:
Let $f : \mathbb{R} \to \mathbb{R}$ be defined by $f(x) = x + 1$, and $g : \mathbb{R} \to \mathbb{R}$ be defined by $g(x) = 2x$.
- Compute $(f \circ g)(x)$:

$$(f \circ g)(x) = f(g(x)) = f(2x) = 2x + 1.$$

- Compute $(g \circ f)(x)$:

$$(g \circ f)(x) = g(f(x)) = g(x+1) = 2(x+1) = 2x + 2.$$

We observe that $(f \circ g)(x) = 2x + 1$ while $(g \circ f)(x) = 2x + 2$. Since the results are different, we conclude that $f \circ g \neq g \circ f$.
This counterexample shows that, in general, function composition is not commutative because changing the order of application produces different results. ∎

8.3 Graphs of Composite Functions

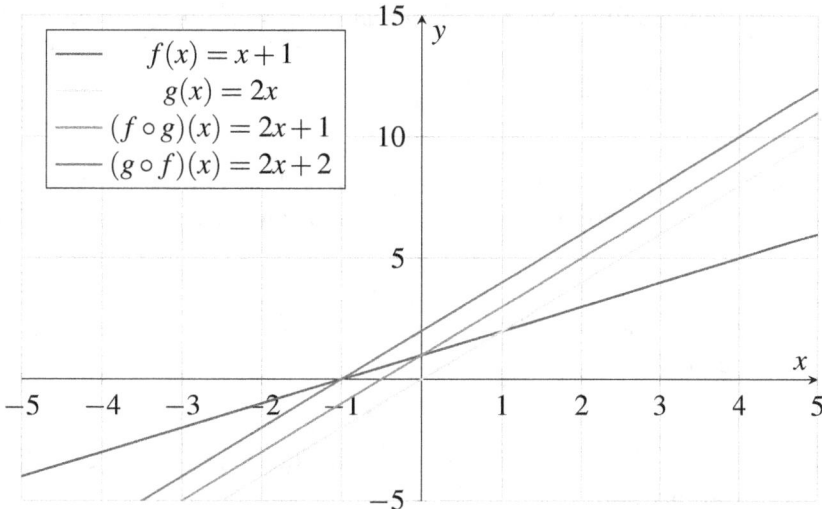

Theorem 8.3.1 Let $f : A \to B$ and $g : B \to C$ be functions. The composition $(g \circ f) : A \to C$ is also a function. Function composition is always associative, i.e., if there are three functions f, g, and h, then $(f \circ (g \circ h)) = ((f \circ g) \circ h)$.

Demostración. To prove that function composition is associative, consider three functions:

$$f : A \to B, \quad g : B \to C, \quad h : C \to D.$$

We want to show that $(f \circ (g \circ h))(x) = ((f \circ g) \circ h)(x)$ for all $x \in A$.
- First, evaluate $(f \circ (g \circ h))(x)$:

$$(f \circ (g \circ h))(x) = f((g \circ h)(x)) = f(g(h(x))).$$

- Now, evaluate $((f \circ g) \circ h)(x)$:

$$((f \circ g) \circ h)(x) = (f \circ g)(h(x)) = f(g(h(x))).$$

We observe that both expressions result in $f(g(h(x)))$. This proves that $(f \circ (g \circ h))(x) = ((f \circ g) \circ h)(x)$ for all $x \in A$.
Thus, function composition is associative. ∎

Corollary 8.3.2 If f and g are bijective functions, then the composition $f \circ g$ is also bijective. In other words, the composition of injective and surjective functions preserves these properties, making it useful to determine the invertibility of composite functions.

Demostración. Given that f and g are bijective functions, we need to prove that $f \circ g$ is injective and surjective.
- **Injectivity**: Suppose $(f \circ g)(x_1) = (f \circ g)(x_2)$ for $x_1, x_2 \in A$. This means $f(g(x_1)) = f(g(x_2))$. Since f is injective, it follows that $g(x_1) = g(x_2)$. And since g is also injective, we conclude that $x_1 = x_2$. Thus, $f \circ g$ is injective.
- **Surjectivity**: Let $y \in C$. Since f is surjective, there exists $b \in B$ such that $f(b) = y$. Since g is surjective, there exists $x \in A$ such that $g(x) = b$. Thus, $(f \circ g)(x) = f(g(x)) = y$. Therefore, $f \circ g$ is surjective.
Since $f \circ g$ is both injective and surjective, it is bijective. ∎

■ **Example 8.14** Consider the functions $f(x) = \sqrt{x}$ and $g(x) = x^2 + 4$. The composition $(f \circ g)(x)$ is:

$$(f \circ g)(x) = f(g(x)) = \sqrt{x^2 + 4}.$$

In this example, the quadratic function is evaluated first, and then the square root is applied to the result. This demonstrates how compositions can be used to construct more complex functions from simpler ones. ■

Exercise 8.13 Given the functions $f(x) = 3x + 1$ and $g(x) = x - 2$, calculate $(f \circ g)(x)$ and $(g \circ f)(x)$. Are they equal?

Exercise 8.14 Let $f(x) = x^3$ and $g(x) = \ln(x)$. Calculate the composition $(g \circ f)(x)$ and analyze the domain of the resulting function.

(R) These exercises highlight the importance of order in composition and how the domain of a composite function depends on the domains of the original functions.

8.3.2 Graph of Composite Functions

Definition 8.3.2 The **graph of a composite function** is obtained by successively applying two functions, f and g, such that the resulting graph corresponds to the composite function $h(x) = f(g(x))$. This composition can be interpreted as applying one transformation on top of another, which can affect both the overall shape of the graph and its location.

■ **Example 8.15** Let $f(x) = x^2$ and $g(x) = x + 2$. The composition $(f \circ g)(x)$ is expressed as $f(g(x)) = (x+2)^2$. To graph this composite function, the displacement in $g(x)$ is performed first, and then the quadratic function is applied. In this case, the graph of $f(g(x))$ will be a parabola shifted 2 units to the left. ■

Lema 8.3.2 If f is a continuous function and g is also continuous, then the composition $f \circ g$ is continuous. This implies that if both original functions have uninterrupted graphs, the graph of the composite function will also be continuous.

Theorem 8.3.3 Let $f : A \to B$ and $g : B \to C$ be two functions such that f and g are differentiable in their respective domains. Then, the composite function $(f \circ g)(x)$ is differentiable, and its derivative is given by the chain rule:

$$(f \circ g)'(x) = f'(g(x)) \cdot g'(x)$$

This property is fundamental for understanding how changes in $g(x)$ affect the graph of the composition.

Corollary 8.3.4 If the function $g(x)$ has a positive derivative in its domain, then the composite function $f(g(x))$ will preserve the direction of growth of f, i.e., if f is increasing, $f \circ g$ will also be increasing. This is useful for determining the slope of the graph of a composite function.

(R) The graph of a composite function can be challenging to interpret without analyzing how the two functions are applied in sequence. The results above are fundamental for understanding

8.3 Graphs of Composite Functions

how to combine transformations and analyze how the resulting graph changes depending on each transformation.

■ **Example 8.16** Consider $f(x) = \sin(x)$ and $g(x) = x^2$. The composition $(f \circ g)(x) = \sin(x^2)$ has a graph that represents oscillation with a constant amplitude, but the frequency changes with the value of x. The function oscillates more rapidly as we move away from the origin due to the quadratic behavior of $g(x)$. ■

Exercise 8.15 Let $f(x) = e^x$ and $g(x) = -x + 1$. Graph the composite function $f(g(x))$ and describe how the original graph of $f(x)$ changes when composed with $g(x)$.

Exercise 8.16 Given the functions $f(x) = \ln(x)$ and $g(x) = x^2 + 1$, graph the function $(f \circ g)(x)$ and analyze the points where the graph is not defined.

(R) These exercises illustrate how compositions affect the graphical properties of functions and allow for visualizing the combined effects of different transformations.

8.3.3 Examples of Composition and Its Applications

■ **Example 8.17** Let $f(x) = x^2$ and $g(x) = 3x + 1$. The composition $(f \circ g)(x)$ is:

$$(f \circ g)(x) = f(g(x)) = (3x + 1)^2 = 9x^2 + 6x + 1$$

In this case, the function f is applied after g, producing a quadratic polynomial. Such compositions are common in models that study quadratic relationships between two linearly related variables. ■

■ **Example 8.18** Consider the functions $f(x) = \sqrt{x}$ and $g(x) = x + 4$. The composition $(f \circ g)(x)$ is given by:

$$(f \circ g)(x) = f(g(x)) = \sqrt{x + 4}$$

The composition of a linear function and a square root function is common in physics applications, such as calculating escape velocity, where a linear variable undergoes a nonlinear transformation to model specific behavior. ■

■ **Example 8.19** Let $f(x) = \sin(x)$ and $g(x) = 2x + \pi$. The composition $(f \circ g)(x)$ is:

$$(f \circ g)(x) = \sin(2x + \pi)$$

This type of composition appears in signal analysis, where $g(x)$ adjusts the phase of a sinusoidal signal, and the function f represents the oscillation. Such adjustments are key to modeling periodic phenomena with variations in frequency and phase. ■

■ **Example 8.20** Consider $f(x) = \ln(x)$ and $g(x) = e^{2x}$. The composition $(f \circ g)(x)$ can be written as:

$$(f \circ g)(x) = \ln(e^{2x}) = 2x$$

In this case, the logarithmic and exponential functions çancel.each other, a property that is very useful in advanced mathematics for solving differential equations and simplifying expressions in complex calculations. ■

■ **Example 8.21** Let $f(x) = |x|$ and $g(x) = x - 3$. The composition $(f \circ g)(x)$ is:

$$(f \circ g)(x) = |x - 3|$$

This composition is frequently used to model distances in geometry and physics. The absolute value ensures that the distance is non-negative, while $g(x)$ shifts the function to center it at a new value. ■

8.4 Solved Exercises

> **Exercise 8.17** Draw the function $f(x) = |x-2|$ and describe how it has been shifted compared to $f(x) = |x|$.

Solution:
The function $f(x) = |x-2|$ is a horizontal shift of the function $f(x) = |x|$. The graph of $f(x) = |x|$ shifts 2 units to the right because the term $x-2$ implies that the absolute value reaches its minimum at $x = 2$.
- **Shift**: The function has shifted **2 units to the right** compared to $f(x) = |x|$.

> **Exercise 8.18** Reflect the function $f(x) = x^2 + 3$ with respect to the x-axis. Write the new equation.

Solution:
To reflect the function with respect to the x-axis, we multiply the entire function by -1:

$$f(x) = x^2 + 3 \quad \rightarrow \quad -f(x) = -(x^2 + 3) = -x^2 - 3$$

The new equation of the reflected function is:

$$f_{\text{reflected}}(x) = -x^2 - 3$$

> **Exercise 8.19** Calculate the vertical scaling of $f(x) = \sin(x)$ by a factor of 3. Represent the new function.

Solution:
The vertical scaling of $f(x) = \sin(x)$ by a factor of 3 is done by multiplying the original function by 3:

$$f_{\text{scaled}}(x) = 3\sin(x)$$

This means the maximum and minimum values of the function are expanded from 1 and -1 to 3 and -3, respectively.

> **Exercise 8.20** Draw the graph of the composite function $f(g(x))$ where $f(x) = \sqrt{x}$ and $g(x) = x - 4$.

Solution:
The composite function $f(g(x))$ is given by:

$$f(g(x)) = f(x-4) = \sqrt{x-4}$$

This is a translation of the function $f(x) = \sqrt{x}$ to the right by 4 units. The graph of the function $f(x) = \sqrt{x}$ starts at the origin, but the function $f(g(x)) = \sqrt{x-4}$ starts at $x = 4$.

> **Exercise 8.21** Determine the resulting function from translating $f(x) = \ln(x)$ two units to the right and three units upward.

Solution:

To translate the function $f(x) = \ln(x)$ two units to the right and three units upward, we make the following changes:

- **Horizontal shift to the right** by 2 units: $f(x) = \ln(x-2)$ - **Vertical shift upward** by 3 units: $f(x) = \ln(x-2)+3$

Thus, the resulting function is:

$$f_{\text{translated}}(x) = \ln(x-2)+3$$

8.5 Proposed Exercises

8.5.1 Translation and Reflection of Functions

Exercise 8.22 Graph the function $f(x) = |x|$ and its translation $f(x-3)$. Describe the effect of the translation.

Exercise 8.23 Reflect the function $f(x) = x^2$ with respect to the y-axis. How does the graph look after the reflection?

Exercise 8.24 Draw the function $f(x) = x^3$ and then its reflection with respect to the x-axis. Write the new equation of the reflected function.

Exercise 8.25 Calculate and graph the vertical translation of $f(x) = \sqrt{x}$ upward by 4 units.

Exercise 8.26 Find the resulting equation of the translation of $f(x) = \ln(x)$ two units to the left and graph it.

8.5.2 Vertical and Horizontal Scaling

Exercise 8.27 Calculate the vertical scaling of $f(x) = x^2$ by a factor of 3. Represent the new function.

Exercise 8.28 Determine the effect of the horizontal scaling of $f(x) = \sin(x)$ by a factor of $1/2$. Write the new function.

Exercise 8.29 Draw the function $f(x) = e^x$ and its vertical scaling by a factor of -2. Describe the effect of this scaling.

Exercise 8.30 Calculate the horizontal scaling of $f(x) = |x|$ by a factor of 4. Write the equation of the new function and graph it.

Exercise 8.31 Graph the function $f(x) = \cos(x)$ and its vertical scaling by a factor of 0,5.

8.5.3 Graphs of Composite Functions

Exercise 8.32 Draw the graph of the composite function $f(g(x))$ where $f(x) = x^2$ and $g(x) = x+2$.

Exercise 8.33 Find the equation of the composite function $f(g(x))$ if $f(x) = \sqrt{x}$ and $g(x) = 3x - 1$. Graph the composite function.

Exercise 8.34 Calculate and graph $h(x) = f(g(x))$ where $f(x) = 2x+1$ and $g(x) = x^2$.

Exercise 8.35 Determine the composite function $f(g(x))$ for $f(x) = \ln(x)$ and $g(x) = e^x$. What is the relationship between f and g?

Exercise 8.36 Graph $f(g(x))$ if $f(x) = |x|$ and $g(x) = x-3$. Describe the transformation applied to f.

9. Composition and Inverse Functions

9.1 Function Composition and Its Notation

9.1.1 Definition of Function Composition

Definition 9.1.1 **Function composition** is an operation that takes two functions, $f : A \to B$ and $g : B \to C$, and produces a new function $h : A \to C$ defined as $h(x) = g(f(x))$. This operation is commonly denoted as $(g \circ f)(x) = g(f(x))$, where f is applied first to the input value, and then g is applied to the result of $f(x)$.

■ **Example 9.1** Let $f(x) = 2x + 3$ and $g(x) = x^2 - 1$. The composition $(g \circ f)(x)$ is found by substituting $f(x)$ into g:

$$(g \circ f)(x) = g(f(x)) = (2x+3)^2 - 1 = 4x^2 + 12x + 8$$

This results in a new function composed of f and g, where f is applied first, followed by g transforming the result. ■

Lema 9.1.1 Function composition is generally not commutative, that is, $f \circ g \neq g \circ f$. In most cases, the order of applying the functions changes the result.

> Theorem 9.1.1 **Associativity of Function Composition:** Let $f : A \to B$, $g : B \to C$, and $h : C \to D$ be functions. The composition of these functions is associative, that is,
>
> $$h \circ (g \circ f) = (h \circ g) \circ f$$
>
> This implies that the order in which the functions are grouped does not affect the final result of the composition.

Corollary 9.1.2 If $f : A \to B$ and $g : B \to C$ are injective functions, then the composition $g \circ f : A \to C$ is also injective. Similarly, if f and g are surjective, then $g \circ f$ will also be surjective. This property ensures that composed functions preserve injectivity and surjectivity, which is fundamental for understanding invertible functions.

> (R) Function compositions allow for sequential transformations to be combined and are frequently used in trajectory calculations, process modeling, and many other areas of applied mathematics.

■ **Example 9.2** Consider the functions $f(x) = \sin(x)$ and $g(x) = 3x$. The composition $(f \circ g)(x)$ is:

$$(f \circ g)(x) = \sin(3x)$$

This type of composition is often used in signal analysis, where the frequency of a sinusoidal signal is adjusted by applying a linear function before the sinusoidal transformation. ■

Exercise 9.1 Given the functions $f(x) = x^2 + 1$ and $g(x) = \sqrt{x-1}$, compute the composition $(g \circ f)(x)$ and determine the domain of the resulting function. ■

Exercise 9.2 Let $f(x) = e^x$ and $g(x) = x^2$. Compute the compositions $(f \circ g)(x)$ and $(g \circ f)(x)$. Analyze whether both functions are equal. ■

> (R) The above exercises help illustrate the importance of the order of composition and how it affects both the value and the domain of the resulting function.

9.1.2 Notation and Properties of Composition

Definition 9.1.2 Function composition is defined as an operation that takes two functions, $f : A \to B$ and $g : B \to C$, and produces a new function $h : A \to C$ defined by $h(x) = (g \circ f)(x) = g(f(x))$. This notation indicates that f is applied to x first, and then g is applied to the result of $f(x)$. The composition is represented with the symbol $\circ$.

■ **Example 9.3** Let $f(x) = x^2$ and $g(x) = 3x + 1$. The composition $(g \circ f)(x)$ is calculated as:

$$(g \circ f)(x) = g(f(x)) = 3x^2 + 1$$

This shows how applying f followed by g transforms the input value, resulting in a new quadratic polynomial. The notation is essential to show the correct sequence of operations. ■

Lema 9.1.2 For any function f, it holds that the composition with the identity function I leaves f unchanged:

$$f \circ I = I \circ f = f$$

where $I(x) = x$ for all x in the domain of f. This property shows that the identity function acts as a neutral element with respect to function composition.

Theorem 9.1.3 Associativity of Composition: If $f : A \to B$, $g : B \to C$, and $h : C \to D$ are functions, then:

$$h \circ (g \circ f) = (h \circ g) \circ f$$

This means that when composing three functions, it does not matter how the compositions are

9.1 Function Composition and Its Notation

grouped; the final result will be the same. This property is important for facilitating the analysis and simplification of multiple chained transformations.

Corollary 9.1.4 If $f : A \to B$ and $g : B \to C$ are bijective functions, then their composition $g \circ f : A \to C$ is also bijective. Additionally, the inverse of the composition is given by:

$$(g \circ f)^{-1} = f^{-1} \circ g^{-1}$$

This result is relevant for ensuring that the combination of two invertible functions remains invertible and for easily calculating the inverse of a composed function.

(R) The notation and properties of composition are fundamental for understanding complex functions, especially when studying successive transformations and analyzing mathematical systems in various contexts.

■ **Example 9.4** Consider the functions $f(x) = \cos(x)$ and $g(x) = 2x + \pi$. The composition $(f \circ g)(x)$ can be written as:

$$(f \circ g)(x) = \cos(2x + \pi)$$

This type of composition is frequently used in physics and engineering to model oscillations or periodic variations with a phase adjustment. ■

Exercise 9.3 Let $f(x) = x^3$ and $g(x) = x + 2$. Compute the composition $(g \circ f)(x)$ and determine whether $(f \circ g)(x)$ is equal to $(g \circ f)(x)$. Explain why or why not.

Exercise 9.4 Given the functions $f(x) = \ln(x)$ and $g(x) = e^x + 1$, calculate $(f \circ g)(x)$ and $(g \circ f)(x)$. Determine the domains of both compositions.

(R) These exercises help illustrate the importance of the order of composition and its effect on the domain and range of the resulting functions.

9.1.3 Examples of Function Composition

■ **Example 9.5** Let $f(x) = 2x + 3$ and $g(x) = x^2$. The composition $(g \circ f)(x)$ is:

$$(g \circ f)(x) = g(f(x)) = (2x + 3)^2 = 4x^2 + 12x + 9$$

In this case, the function f is applied first, followed by g on the result of $f(x)$. This composition results in a second-degree polynomial. ■

■ **Example 9.6** Consider the functions $f(x) = \sin(x)$ and $g(x) = 3x$. The composition $(f \circ g)(x)$ is:

$$(f \circ g)(x) = \sin(3x)$$

This type of composition is common in signal analysis, where a linear transformation is applied to the input before passing it through a sinusoidal function. ■

■ **Example 9.7** Let $f(x) = |x|$ and $g(x) = x - 2$. The composition $(f \circ g)(x)$ is calculated as:

$$(f \circ g)(x) = |x - 2|$$

This example shows how a function can be modified to ensure that all output values are non-negative, which is useful in applications involving distances. ■

Capítulo 9. Composition and Inverse Functions

■ **Example 9.8** Let $f(x) = e^x$ and $g(x) = -x$. The composition $(f \circ g)(x)$ is:

$$(f \circ g)(x) = e^{-x}$$

This type of composition arises in the study of exponential decay, as seen in models of radioactive decay or cooling of a body. ■

■ **Example 9.9** Let $f(x) = \ln(x)$ and $g(x) = x^2 + 1$. The composition $(f \circ g)(x)$ is:

$$(f \circ g)(x) = \ln(x^2 + 1)$$

This composition is commonly used when transforming a quadratic function with a logarithmic function, especially in contexts involving logarithmic scales. ■

■ **Example 9.10** Consider $f(x) = x^3$ and $g(x) = \sqrt{x}$. The composition $(f \circ g)(x)$ is:

$$(f \circ g)(x) = (\sqrt{x})^3 = x^{3/2}$$

In this case, the square root function is applied first, followed by cubing the result. This composition is common in problems requiring simplification of expressions involving powers and roots. ■

9.2 Inverse of a Function and Its Calculation

9.2.1 Definition of an Inverse Function

Definition 9.2.1 Given a function $f : A \to B$, f is said to have an **inverse function** $f^{-1} : B \to A$ if, for each element $b \in B$, there exists a unique element $a \in A$ such that $f(a) = b$. The inverse function f^{-1} is defined such that:

$$f^{-1}(f(x)) = x \quad \text{and} \quad f(f^{-1}(y)) = y$$

for all values $x \in A$ and $y \in B$. This means that the inverse function ündoes"the application of the original function.

■ **Example 9.11** Consider the function $f(x) = 2x + 3$, which is a linear function. To find its inverse f^{-1}, first set $y = 2x + 3$ and solve for x:

$$y = 2x + 3 \implies x = \frac{y-3}{2}$$

Thus, the inverse function is $f^{-1}(y) = \frac{y-3}{2}$. In this case, f and f^{-1} satisfy:

$$f^{-1}(f(x)) = x \quad \text{and} \quad f(f^{-1}(y)) = y$$

This confirms that f and f^{-1} are inverse functions. ■

9.2 Inverse of a Function and Its Calculation

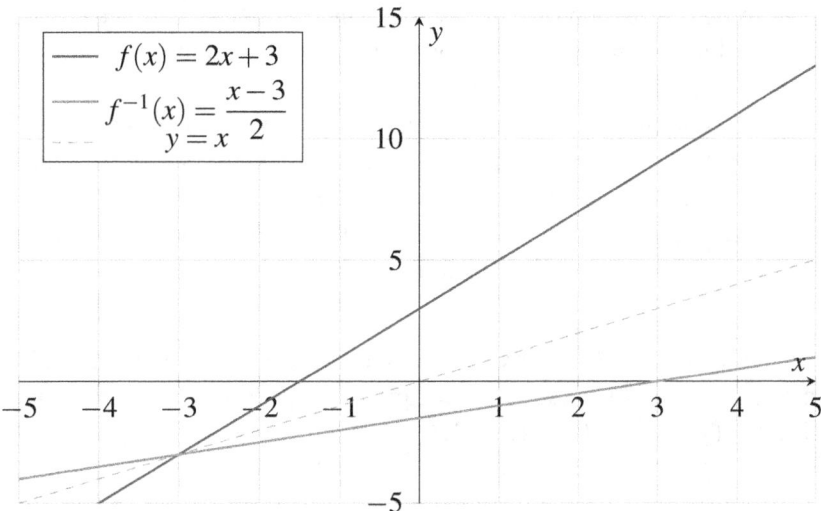

■ **Example 9.12** Consider the function $f(x) = x^3$, which is a cubic function. To find its inverse f^{-1}, set $y = x^3$ and solve for x:

$$y = x^3 \implies x = \sqrt[3]{y}$$

Thus, the inverse function is $f^{-1}(y) = \sqrt[3]{y}$. In terms of x, we can write $f^{-1}(x) = \sqrt[3]{x}$. In this case, f and f^{-1} satisfy:

$$f^{-1}(f(x)) = x \quad \text{and} \quad f(f^{-1}(x)) = x$$

This confirms that f and f^{-1} are inverse functions. ■

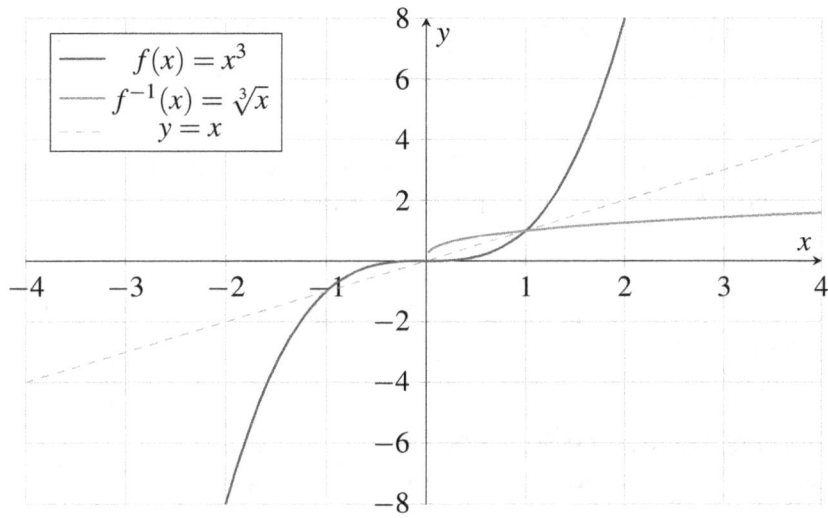

Lema 9.2.1 A function $f : A \to B$ has an inverse if and only if it is bijective, meaning f must be injective (each value in the range has a unique preimage) and surjective (every element in B has a preimage in A).

Proof:
First, we prove that if $f : A \to B$ is bijective, then it has an inverse.
1. **Sufficiency (If bijective, then it has an inverse):** - If f is injective, for each $b \in \text{Im}(f)$, there exists a unique $a \in A$ such that $f(a) = b$. - If f is surjective, then $\text{Im}(f) = B$, meaning for every

$b \in B$, there exists some $a \in A$ such that $f(a) = b$. - Since f is bijective, we can define an inverse function $f^{-1}: B \to A$ that assigns each $b \in B$ the unique $a \in A$ such that $f(a) = b$. Therefore, f has an inverse.

2. **Necessity (If it has an inverse, then it is bijective):** - If f has an inverse $f^{-1}: B \to A$, then for every $b \in B$, $f^{-1}(b) = a$ such that $f(a) = b$. This implies f is surjective, as for every $b \in B$, there exists $a \in A$ mapping to b. - Additionally, if $f(a_1) = f(a_2)$, applying f^{-1} to both sides yields $f^{-1}(f(a_1)) = f^{-1}(f(a_2))$, which implies $a_1 = a_2$. Thus, f is injective. - Therefore, f is bijective. Hence, f has an inverse if and only if it is bijective.

□

> **Theorem 9.2.1 Uniqueness of the Inverse Function:** If a function $f : A \to B$ has an inverse function, then that inverse is unique. This means there cannot exist more than one function $g : B \to A$ such that $g(f(x)) = x$ for all $x \in A$.

Proof:
Suppose $f : A \to B$ has an inverse and there exist two functions $g : B \to A$ and $h : B \to A$ such that:

$$g(f(x)) = x \quad \text{and} \quad h(f(x)) = x \quad \text{for all } x \in A.$$

To prove $g = h$:
Let $y \in B$. Since f is surjective, there exists $x \in A$ such that $f(x) = y$.
Applying g and h to $y = f(x)$:

$$g(y) = g(f(x)) = x \quad \text{and} \quad h(y) = h(f(x)) = x.$$

Thus, $g(y) = h(y)$ for all $y \in B$, implying $g = h$.
Hence, the inverse of f is unique.

□

> **Exercise 9.5** Determine whether the function $f(x) = x^3 - 2$ has an inverse. If it does, find $f^{-1}(x)$.

> **Exercise 9.6** Let $f(x) = \frac{1}{x}$ with $x \neq 0$. Find the inverse function $f^{-1}(x)$ and verify that $f(f^{-1}(x)) = x$ and $f^{-1}(f(x)) = x$ for all permissible values.

> (R) The above exercises illustrate how to find the inverse function and verify its validity using the composition property.

9.2.2 Method for Finding the Inverse Function

> **Definition 9.2.2** To find the **inverse function** of a function $f(x)$, the following general method is applied:
> 1. Set $y = f(x)$. 2. Solve for x in terms of y. 3. Interchange x and y to obtain the inverse function $f^{-1}(x)$.
>
> This process is applicable as long as $f(x)$ is bijective, meaning injective and surjective, which guarantees the existence of a unique inverse function.

> ■ **Example 9.13** Consider the function $f(x) = \frac{3x-1}{2}$. We want to find the inverse function $f^{-1}(x)$.
>
> $$y = \frac{3x-1}{2}$$

9.2 Inverse of a Function and Its Calculation

To solve for x, first multiply both sides by 2:

$$2y = 3x - 1$$

Next, add 1 to both sides:

$$3x = 2y + 1$$

Finally, divide by 3:

$$x = \frac{2y+1}{3}$$

Interchanging x and y, we get:

$$f^{-1}(x) = \frac{2x+1}{3}$$

Thus, the inverse function of $f(x) = \frac{3x-1}{2}$ is $f^{-1}(x) = \frac{2x+1}{3}$. ∎

Lema 9.2.2 If a function f is injective, then for each output value y, there exists a unique input value x such that $f(x) = y$. This is essential for being able to solve for x from the equation $y = f(x)$ and thus find the inverse.

Demostración. Suppose $f : A \to B$ is an injective function. By definition of injectivity, if $f(x_1) = f(x_2)$, then $x_1 = x_2$. This means that for each output value $y \in \text{Im}(f)$, there exists at most one input value $x \in A$ such that $f(x) = y$.

Now, given $y \in \text{Im}(f)$, we know there exists some $x \in A$ such that $f(x) = y$ (by the definition of image). Furthermore, the injectivity of f guarantees that this value x is unique.

Therefore, for each output value y, there exists a unique input value x such that $f(x) = y$. This allows us to solve for x in terms of y and define the inverse function of f. ∎

> **Theorem 9.2.2 Criterion for the Existence of an Inverse:** A function $f : A \to B$ has an inverse function if and only if it is bijective. That is, it must be injective (every value in the image has a unique preimage) and surjective (all of B is covered by the function f).

Demostración. We will prove both directions of the statement.
1. **Sufficiency (If f is bijective, then it has an inverse):** - Suppose $f : A \to B$ is bijective, meaning it is injective and surjective. - **Injectivity** ensures that for each $y \in B$, there is at most one $x \in A$ such that $f(x) = y$. - **Surjectivity** ensures that for each $y \in B$, there exists at least one $x \in A$ such that $f(x) = y$. - Combining both properties, for each $y \in B$, there is exactly one $x \in A$ such that $f(x) = y$. Therefore, we can define an inverse function $f^{-1} : B \to A$ that assigns each $y \in B$ its corresponding $x \in A$, satisfying $f^{-1}(f(x)) = x$ and $f(f^{-1}(y)) = y$. This shows that f has an inverse.
2. **Necessity (If f has an inverse, then it is bijective):** - Suppose f has an inverse $f^{-1} : B \to A$. - The existence of f^{-1} implies that for each $y \in B$, there exists an $x \in A$ such that $f(x) = y$, showing that f is surjective. - Additionally, if $f(x_1) = f(x_2)$, applying f^{-1} to both sides gives $f^{-1}(f(x_1)) = f^{-1}(f(x_2))$, which implies $x_1 = x_2$, thus f is injective.

Hence, f has an inverse if and only if it is bijective. ∎

Corollary 9.2.3 If a function f is strictly increasing or strictly decreasing on its domain, then f has an inverse function. This is because strict monotonicity guarantees injectivity.

Demostración. Suppose $f : A \to B$ is strictly increasing or strictly decreasing on its domain.
1. **If f is strictly increasing:** For any $x_1, x_2 \in A$ with $x_1 < x_2$, we have $f(x_1) < f(x_2)$. This implies that no two different values of x can map to the same image, i.e., f is injective.
2. **If f is strictly decreasing:** For any $x_1, x_2 \in A$ with $x_1 < x_2$, we have $f(x_1) > f(x_2)$. Again, no two different values of x can map to the same image, so f is injective.
Since f is injective in both cases, there exists an inverse function $f^{-1} : B \to A$ such that $f^{-1}(f(x)) = x$ for all $x \in A$. Strict monotonicity ensures injectivity, and thus f has an inverse.
□

■

> R The process of finding the inverse requires that the function is bijective, which guarantees that each output value is associated with a unique input value. It is important to verify that the function is injective before attempting to find the inverse.

■ **Example 9.14** Let $f(x) = x^2 + 1$ with the domain restricted to $x \geq 0$. To find the inverse function, follow these steps:
1. Set $y = x^2 + 1$. 2. Solve for x:

$$y - 1 = x^2$$

$$x = \sqrt{y - 1}$$

3. Interchange x and y to obtain:

$$f^{-1}(x) = \sqrt{x - 1}$$

The restriction $x \geq 0$ on the domain of f ensures that f is injective and thus has an inverse function.

■

> **Exercise 9.7** Determine whether the function $f(x) = 5 - 2x$ has an inverse, and if so, find it. ■

> **Exercise 9.8** Let $f(x) = \frac{1}{x}$ with $x > 0$. Find the inverse function $f^{-1}(x)$ and verify that $f(f^{-1}(x)) = x$ and $f^{-1}(f(x)) = x$. ■

> R The presented exercises demonstrate how to apply the method for finding the inverse function and emphasize the importance of verifying that the original function is injective to ensure the existence of the inverse.

9.2.3 Verification of the Inverse

> **Definition 9.2.3** The **verification of an inverse function** involves demonstrating that a function f and its proposed inverse f^{-1} satisfy the following two properties:
> 1. $f^{-1}(f(x)) = x$ for all x in the domain of f. 2. $f(f^{-1}(y)) = y$ for all y in the domain of f^{-1}.
> These properties ensure that f and f^{-1} mutually cancel each other, meaning the inverse function undoes the effect of the original function.

9.2 Inverse of a Function and Its Calculation

■ **Example 9.15** Consider the function $f(x) = 2x+3$ and its proposed inverse $f^{-1}(y) = \frac{y-3}{2}$. We will verify whether they are indeed inverses.
First, calculate $f^{-1}(f(x))$:

$$f^{-1}(f(x)) = f^{-1}(2x+3) = \frac{(2x+3)-3}{2} = x$$

Next, calculate $f(f^{-1}(y))$:

$$f(f^{-1}(y)) = f\left(\frac{y-3}{2}\right) = 2 \cdot \frac{y-3}{2} + 3 = y - 3 + 3 = y$$

Since both properties are satisfied, we conclude that $f(x) = 2x+3$ and $f^{-1}(y) = \frac{y-3}{2}$ are inverse functions. ■

Lema 9.2.3 If a function f has an inverse f^{-1}, then the compositions $f^{-1}(f(x)) = x$ and $f(f^{-1}(y)) = y$ must hold for all values in the domains of f and f^{-1}. This implies that the function is both injective and surjective.

Demostración. Suppose $f : A \to B$ is a function with an inverse $f^{-1} : B \to A$. By the definition of an inverse function, the composition must satisfy the following properties:
1. $f^{-1}(f(x)) = x$ for all $x \in A$. 2. $f(f^{-1}(y)) = y$ for all $y \in B$.
These equalities ensure that f is injective and surjective:
- **Injectivity**: If $f(x_1) = f(x_2)$, then applying f^{-1} to both sides gives $f^{-1}(f(x_1)) = f^{-1}(f(x_2))$, which implies $x_1 = x_2$. Thus, f is injective.
- **Surjectivity**: For any $y \in B$, there exists an $x \in A$ such that $f(x) = y$ (because $f(f^{-1}(y)) = y$ for all $y \in B$), proving that f is surjective.
Therefore, f is bijective, confirming the existence of a unique inverse function. ∎

> **Theorem 9.2.4 Criterion for Verifying the Inverse:** Let $f : A \to B$ be a function and let $g : B \to A$ such that $f(g(y)) = y$ for all $y \in B$ and $g(f(x)) = x$ for all $x \in A$. Then $g = f^{-1}$ and $f = g^{-1}$. This result guarantees the uniqueness of the inverse function if both compositions yield the identity function.

Demostración. Suppose $f : A \to B$ is a function and $g : B \to A$ satisfies:
1. $f(g(y)) = y$ for all $y \in B$. 2. $g(f(x)) = x$ for all $x \in A$.
These two properties imply that g is the inverse of f and f is the inverse of g because:
- The condition $f(g(y)) = y$ for all $y \in B$ means $f \circ g = \text{id}_B$, where id_B is the identity function on B.
- The condition $g(f(x)) = x$ for all $x \in A$ means $g \circ f = \text{id}_A$, where id_A is the identity function on A.
Since both compositions return the identity function on their respective sets, g is the inverse of f, i.e., $g = f^{-1}$ and $f = g^{-1}$. ∎

> **Corollary 9.2.5** If a function f is strictly monotonic (i.e., strictly increasing or strictly decreasing) on its domain, then it has an inverse function that can be verified by applying the criterion for verifying the inverse.

Demostración. If $f : A \to B$ is strictly monotonic on its domain (strictly increasing or strictly decreasing), then f is injective because strict monotonicity ensures that no two distinct input values produce the same output.
Since f is injective and its image covers B, it is also surjective onto B. Therefore, f is bijective, meaning it has an inverse function $f^{-1} : B \to A$.

To verify that f^{-1} is indeed the inverse of f, we can apply the criterion for verifying the inverse: if $f(f^{-1}(y)) = y$ for all $y \in B$ and $f^{-1}(f(x)) = x$ for all $x \in A$, then f^{-1} satisfies the properties of the inverse function of f.

Thus, the strict monotonicity of f guarantees the existence and uniqueness of its inverse f^{-1}. ∎

> (R) Verifying the inverse is a crucial step to confirm that a function and its inverse undo each other, which is a fundamental property in function theory and ensures the mathematical consistency of related calculations.

■ **Example 9.16** Let $f(x) = x^3 + 1$ and its proposed inverse $f^{-1}(y) = \sqrt[3]{y-1}$. Verify both properties:

1. $f^{-1}(f(x)) = f^{-1}(x^3 + 1) = \sqrt[3]{(x^3+1) - 1} = \sqrt[3]{x^3} = x$ 2. $f(f^{-1}(y)) = f(\sqrt[3]{y-1}) = (\sqrt[3]{y-1})^3 + 1 = y - 1 + 1 = y$

Since both properties hold, f and f^{-1} are inverse functions. ∎

> Exercise 9.9 Verify that the function $f(x) = 4x - 5$ and its inverse $f^{-1}(y) = \frac{y+5}{4}$ satisfy the properties of inverse functions.

> Exercise 9.10 Let $f(x) = \frac{x+2}{3}$. Find the inverse function and verify that $f(f^{-1}(y)) = y$ and $f^{-1}(f(x)) = x$.

> (R) The exercises demonstrate how to apply the method for verifying the inverse function, emphasizing the importance of the function being bijective to guarantee the existence of a unique inverse.

9.3 Properties of Inverse Functions

9.3.1 Relationship Between a Function and Its Inverse

Definition 9.3.1 Let $f : A \to B$ be a bijective function. The **inverse function** of f, denoted as $f^{-1} : B \to A$, is such that for every $y \in B$, $f^{-1}(y) = x$ if and only if $f(x) = y$. This implies that $f(f^{-1}(y)) = y$ and $f^{-1}(f(x)) = x$ for all $x \in A$ and $y \in B$.

> (R) The fundamental property of an inverse function is that it undoes the action of the original function. This means applying f and then f^{-1} returns the initial value, and vice versa.

■ **Example 9.17** Consider the function $f(x) = 3x - 5$ and its inverse $f^{-1}(x) = \dfrac{x+5}{3}$. Let us verify the relationship between them:

$$f(f^{-1}(x)) = f\left(\frac{x+5}{3}\right) = 3 \cdot \frac{x+5}{3} - 5 = x + 5 - 5 = x$$

$$f^{-1}(f(x)) = f^{-1}(3x - 5) = \frac{3x - 5 + 5}{3} = \frac{3x}{3} = x$$

These relationships confirm that f and f^{-1} are inverse functions. ∎

Lema 9.3.1 If f is a strictly monotonic function, then f has an inverse function f^{-1}, and both are bijective functions.

9.3 Properties of Inverse Functions

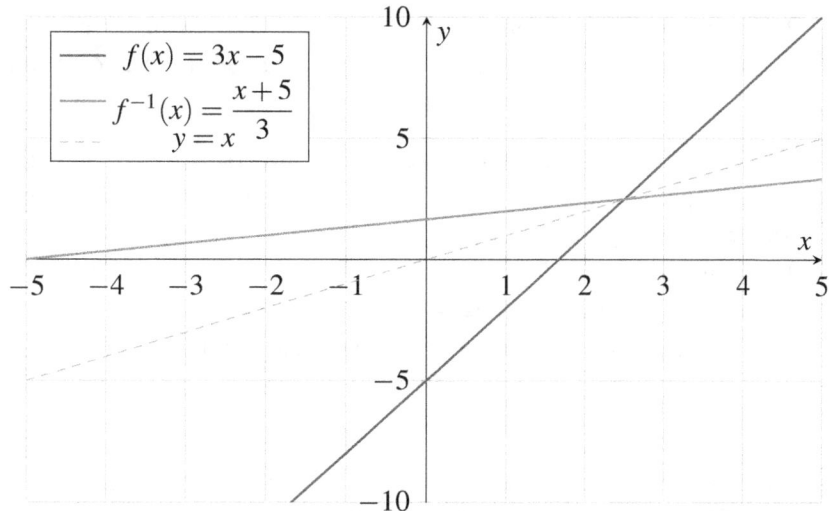

Figura 9.3.1: *Graph of $f(x) = 3x - 5$ and its inverse $f^{-1}(x) = \dfrac{x+5}{3}$*

Demostración. Let $f : A \to B$ be a strictly monotonic function on its domain (i.e., strictly increasing or strictly decreasing).
1. **Injectivity**: The strict monotonicity of f implies that no two distinct input values produce the same output. That is, if $f(x_1) = f(x_2)$, then $x_1 = x_2$. This guarantees that f is injective.
2. **Surjectivity**: Suppose $B = \text{Im}(f)$, the set of values attained by f. If f is strictly monotonic and continuous over a closed interval, then f covers the entire range B without gaps, making f surjective onto B.

Since f is both injective and surjective, it is bijective, and therefore has an inverse $f^{-1} : B \to A$. Finally, f^{-1} is also strictly monotonic (if f is strictly increasing, f^{-1} is increasing; if f is strictly decreasing, f^{-1} is decreasing), ensuring that f^{-1} is also injective and surjective. Hence, both f and f^{-1} are bijective functions. ∎

> **Theorem 9.3.1 Relationship Between a Function and Its Inverse:** Let $f : A \to B$ be a bijective function and let $f^{-1} : B \to A$ be its inverse. Then the following properties hold:
> 1. $f(f^{-1}(y)) = y$ for all $y \in B$.
> 2. $f^{-1}(f(x)) = x$ for all $x \in A$.
>
> These properties ensure that the composition of a function and its inverse always returns the identity.

Demostración. Since $f : A \to B$ is bijective, there exists an inverse function $f^{-1} : B \to A$ such that the compositions $f \circ f^{-1}$ and $f^{-1} \circ f$ return the identity on their respective domains. Let us prove both properties:
1. **Proof of $f(f^{-1}(y)) = y$ for all $y \in B$:** - For any $y \in B$, by the definition of the inverse, there exists $x \in A$ such that $f(x) = y$. - Applying f to $f^{-1}(y)$, we get $f(f^{-1}(y)) = y$, which shows that $f \circ f^{-1} = \text{id}_B$, the identity on B.
2. **Proof of $f^{-1}(f(x)) = x$ for all $x \in A$:** - For any $x \in A$, applying f^{-1} to $f(x)$, we get $f^{-1}(f(x)) = x$, which shows that $f^{-1} \circ f = \text{id}_A$, the identity on A.

These properties confirm that the composition of a bijective function and its inverse returns the identity in both directions, satisfying the relationship between f and f^{-1}. ∎

Corollary 9.3.2 The graph of a function f and its inverse f^{-1} are symmetric with respect to the line $y = x$. This is because the function and its inverse exchange the roles of x and y.

Demostración. Let $f : A \to B$ be a bijective function with inverse $f^{-1} : B \to A$. By definition, if (a,b) is a point on the graph of f, this means $f(a) = b$. On the other hand, (b,a) will be a point on the graph of f^{-1} because $f^{-1}(b) = a$.

We observe that the points (a,b) and (b,a) are symmetric with respect to the line $y = x$, as this line acts as the axis of symmetry when interchanging the coordinates x and y.

Hence, the graphs of f and f^{-1} are symmetric with respect to the line $y = x$. ∎

> This corollary is particularly useful when working with graphs, as it provides an intuitive way to visualize the relationship between a function and its inverse.

■ **Example 9.18** Let $f(x) = x^2$, with the domain restricted to $[0, \infty)$. The inverse function is $f^{-1}(y) = \sqrt{y}$, with the domain $[0, \infty)$. Both functions are symmetric with respect to the line $y = x$, consistent with the corollary. ■

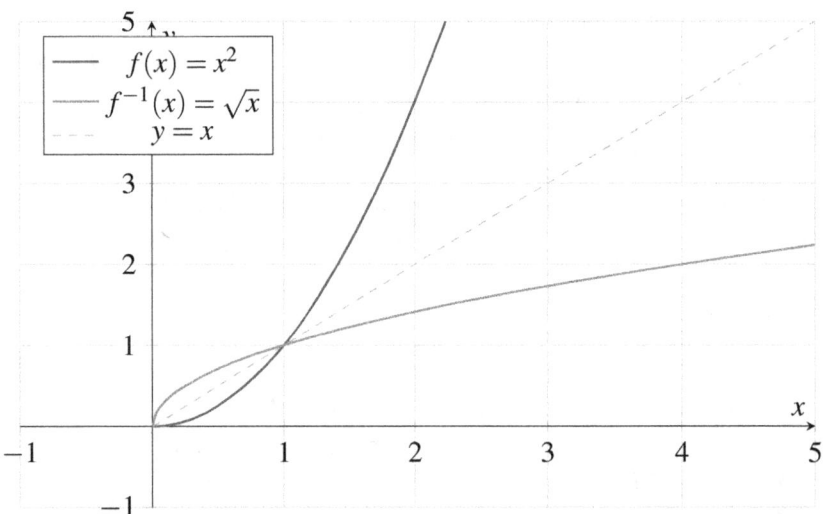

Figura 9.3.2: *Graph of $f(x) = x^2$ and its inverse $f^{-1}(x) = \sqrt{x}$*

Exercise 9.11 Verify that $f(x) = 5x + 2$ and its inverse $f^{-1}(y) = \frac{y-2}{5}$ satisfy the property $f(f^{-1}(y)) = y$ and $f^{-1}(f(x)) = x$. ■

Exercise 9.12 Let $f(x) = 2x - 3$. Find its inverse and verify that the graphs of f and its inverse are symmetric with respect to the line $y = x$. ■

> The properties of inverse functions have important applications in various areas of mathematics, such as analysis and equation solving. Understanding the relationship between a function and its inverse is key to many algebraic and geometric methods.

9.3 Properties of Inverse Functions

9.3.2 Graph of an Inverse Function

Definition 9.3.2 The **graph of the inverse function** of a function f is the reflection of the graph of f with respect to the line $y = x$. This means that if (a,b) is a point on the graph of f, then (b,a) is a point on the graph of f^{-1}.

(R) To visualize the relationship between a function and its inverse, it is helpful to plot both on the same coordinate system. This makes it easier to observe their symmetry with respect to the line $y = x$.

■ **Example 9.19** Consider the function $f(x) = 2x + 3$ and its inverse $f^{-1}(y) = \frac{y-3}{2}$. To graph both functions:
- The graph of f is a straight line with a slope of 2 and a y-intercept of 3. - The graph of f^{-1} is a straight line with a slope of $\frac{1}{2}$ and a y-intercept of $-\frac{3}{2}$.
We observe that both graphs are symmetric with respect to the line $y = x$. ■

Lema 9.3.2 If f is a bijective function, then the graph of f^{-1} can be obtained by reflecting the graph of f with respect to the line $y = x$. Moreover, the intersection of the graph of f and the graph of f^{-1} occurs at points lying on $y = x$.

Theorem 9.3.3 **Symmetry of the Graph of a Function and Its Inverse:** The graph of a function f and the graph of its inverse f^{-1} are symmetric with respect to the line $y = x$. Specifically, if $f(a) = b$, then $f^{-1}(b) = a$, and these points are exchanged under the reflection.

Corollary 9.3.4 If the graph of a function f intersects the line $y = x$, then the graph of f^{-1} also intersects this line at the same points. This happens because such points are of the form (a,a), and thus belong to both functions.

■ **Example 9.20** Let the function $f(x) = x^3$ and its inverse $f^{-1}(y) = \sqrt[3]{y}$. The graph of f and f^{-1} intersect the line $y = x$ and are symmetric with respect to it. In this case, the symmetry is easy to observe due to the cubic nature of the function. ■

Exercise 9.13 Graph the function $f(x) = 3x - 2$ and its inverse $f^{-1}(y) = \frac{y+2}{3}$. Verify that both graphs are symmetric with respect to the line $y = x$.

Exercise 9.14 Consider the function $f(x) = \frac{1}{x}$, with domain $x \neq 0$. Graph f and its inverse, and demonstrate that both graphs are symmetric with respect to the line $y = x$.

(R) Understanding the graphical relationship between a function and its inverse is fundamental for many advanced topics in mathematics, including calculus and functional analysis. The symmetry with respect to $y = x$ provides a powerful visual way to understand how a function and its inverse interact.

9.3.3 Applications of Inverse Functions

Definition 9.3.3 An **inverse function** is used to reverse the effect of a given function. It is especially useful in applications where we want to find the original input value corresponding to a specific output. For example, if $f(x)$ represents the distance traveled as a function of time, the inverse $f^{-1}(y)$ can be used to determine the time required to travel a certain distance.

Inverse functions are found in many real-world contexts, such as temperature conversion between Celsius and Fahrenheit, or calculating the intensity of a signal relative to its distance from the source. These applications allow us to model and solve problems where reversing a relationship is required.

■ **Example 9.21** Consider the function $f(x) = 2x + 3$, which represents the price of a product based on a fixed amount plus a proportional cost. The inverse function $f^{-1}(y) = \frac{y-3}{2}$ allows us to determine the base cost given the total price, which is useful in financial calculations and audits. ■

Lema 9.3.3 For a function f and its inverse f^{-1}, if $f(a) = b$, then $f^{-1}(b) = a$. This property allows us to interchange input and output values, which is fundamental for solving problems requiring an inverse function.

Theorem 9.3.5 **Inverse Relationship and Composition:** If f is a function with inverse f^{-1}, then $f(f^{-1}(x)) = x$ and $f^{-1}(f(x)) = x$ for any value x in the respective domains. This relationship shows that the composition of a function and its inverse always yields the original value.

Proof:
Since f has an inverse f^{-1}, by the definition of an inverse function, for any $x \in B$ and $y \in A$, the following properties hold:
1. $f^{-1}(f(x)) = x$ for all $x \in A$.
- This means that applying f to x yields a value in B, and applying f^{-1} to this value returns the original x.
2. $f(f^{-1}(y)) = y$ for all $y \in B$.
- This means that applying f^{-1} to y yields a value in A, and applying f to this value returns the original y.
These properties define the relationship between a function and its inverse: the composition of f with f^{-1} (and vice versa) returns the original value.
□

Corollary 9.3.6 If f and f^{-1} are inverses, then their graphs are symmetric with respect to the line $y = x$. This property can be used to graphically analyze how a function and its inverse relate and intersect.

■ **Example 9.22** A practical application of inverse functions is the conversion of temperatures between Celsius and Fahrenheit. The relationship between Fahrenheit (F) and Celsius (C) is given by $F = \frac{9}{5}C + 32$. The inverse, converting from Fahrenheit to Celsius, is $C = \frac{5}{9}(F - 32)$. This inverse function allows us to convert between scales, aiding in understanding temperatures in different contexts. ■

Exercise 9.15 Determine the inverse function of $f(x) = 5x - 7$ and use the composition property to verify that $f(f^{-1}(x)) = x$.

Exercise 9.16 Let the function $f(x) = e^x$. Find its inverse and determine the value of the inverse function at $y = 2$. Explain the significance of this value in the context of exponential growth. ■

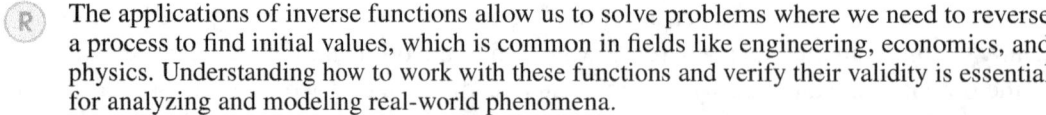 The applications of inverse functions allow us to solve problems where we need to reverse a process to find initial values, which is common in fields like engineering, economics, and physics. Understanding how to work with these functions and verify their validity is essential for analyzing and modeling real-world phenomena.

9.4 Solved Exercises

Exercise 9.17 Calculate the composition $(f \circ g)(x)$ for $f(x) = x^2$ and $g(x) = 2x+1$.

Solution:
The composition $(f \circ g)(x)$ is obtained by substituting the function $g(x)$ into the function $f(x)$:

$$(f \circ g)(x) = f(g(x)) = f(2x+1) = (2x+1)^2 = 4x^2 + 4x + 1$$

Exercise 9.18 Find the inverse of $f(x) = \frac{x-3}{2}$ and verify if $f(f^{-1}(x)) = x$.

Solution:
To find the inverse $f^{-1}(x)$, swap x and y and solve for y:

$$y = \frac{x-3}{2}$$

Swap x and y:

$$x = \frac{y-3}{2}$$

Multiply both sides by 2:

$$2x = y - 3$$

Add 3 to both sides:

$$y = 2x + 3$$

Thus, the inverse function is:

$$f^{-1}(x) = 2x + 3$$

Verification:

$$f(f^{-1}(x)) = f(2x+3) = \frac{(2x+3)-3}{2} = \frac{2x}{2} = x$$

Exercise 9.19 Find the composition $(g \circ f)(x)$ where $f(x) = \sin(x)$ and $g(x) = x+2$. Is $(f \circ g)(x)$ equal to $(g \circ f)(x)$?

Solution:
First, calculate $(g \circ f)(x)$:

$$(g \circ f)(x) = g(f(x)) = g(\sin(x)) = \sin(x) + 2$$

Now, calculate $(f \circ g)(x)$:

$$(f \circ g)(x) = f(g(x)) = f(x+2) = \sin(x+2)$$

Comparison:

$$(g \circ f)(x) = \sin(x) + 2 \quad \text{and} \quad (f \circ g)(x) = \sin(x+2)$$

Therefore, $(f \circ g)(x) \neq (g \circ f)(x)$.

> **Exercise 9.20** Find the inverse of the function $f(x) = x^3 + 1$ and determine its domain and range.

Solution:
To find the inverse $f^{-1}(x)$, swap x and y and solve for y:

$$y = x^3 + 1$$

Swap x and y:

$$x = y^3 + 1$$

Subtract 1 from both sides:

$$x - 1 = y^3$$

Take the cube root:

$$y = \sqrt[3]{x-1}$$

Thus, the inverse function is:

$$f^{-1}(x) = \sqrt[3]{x-1}$$

Domain and Range:
- **Domain** of $f(x) = x^3 + 1$: All real numbers $\mathbb{R}$. - **Range** of $f(x) = x^3 + 1$: All real numbers $\mathbb{R}$.
The inverse function $f^{-1}(x) = \sqrt[3]{x-1}$ also has domain and range on all real numbers $\mathbb{R}$.

> **Exercise 9.21** Verify that $f(x) = 3x - 5$ and $g(x) = \frac{x+5}{3}$ are inverses of each other.

Solution:
First, calculate $f(g(x))$:

$$f(g(x)) = f\left(\frac{x+5}{3}\right) = 3 \cdot \frac{x+5}{3} - 5 = x + 5 - 5 = x$$

Now, calculate $g(f(x))$:

$$g(f(x)) = g(3x-5) = \frac{(3x-5)+5}{3} = \frac{3x}{3} = x$$

Since both compositions result in x, we can conclude that $f(x)$ and $g(x)$ are inverses of each other.

9.5 Proposed Exercises

9.5.1 Function Composition and Notation

Exercise 9.22 Calculate the composition $(f \circ g)(x)$ for $f(x) = 3x - 4$ and $g(x) = x^2 + 2$.

Exercise 9.23 Determine the composed function $(g \circ f)(x)$ if $f(x) = \sin(x)$ and $g(x) = x^3$. Write the resulting equation.

Exercise 9.24 Find $(f \circ g)(2)$ for $f(x) = x + 1$ and $g(x) = 2x^2 - 3$.

Exercise 9.25 Calculate the composition $(f \circ g)(x)$ where $f(x) = \ln(x)$ and $g(x) = e^{2x}$. Simplify the result.

Exercise 9.26 Determine if the composition $(f \circ g)(x)$ is equal to $(g \circ f)(x)$ for $f(x) = x + 5$ and $g(x) = 3x$.

9.5.2 Inverse of a Function and Its Calculation

Exercise 9.27 Find the inverse of the function $f(x) = 2x + 3$. Verify if $f(f^{-1}(x)) = x$.

Exercise 9.28 Determine the inverse of $f(x) = \frac{1}{x-1}$ and calculate $f^{-1}(2)$.

Exercise 9.29 Find the inverse of $f(x) = x^2 - 4$, restricted to the domain $x \geq 0$.

Exercise 9.30 Calculate the inverse of $f(x) = \sqrt{x-3}$ and represent its graph.

Exercise 9.31 Verify that $f(x) = 5x - 7$ and $g(x) = \frac{x+7}{5}$ are inverse functions.

9.5.3 Properties of Inverse Functions

Exercise 9.32 Verify if $f(x) = 3x - 2$ and $g(x) = \frac{x+2}{3}$ satisfy $f(g(x)) = x$ and $g(f(x)) = x$.

Exercise 9.33 Determine if the function $f(x) = x^3 + 1$ has an inverse. If so, calculate $f^{-1}(x)$.

Exercise 9.34 Find the inverse of $f(x) = \frac{2x-3}{5}$ and demonstrate the relationship between f and f^{-1}.

Exercise 9.35 Verify graphically that the function $f(x) = \ln(x)$ and its inverse $f^{-1}(x) = e^x$ are reflections with respect to the line $y = x$.

Exercise 9.36 Calculate the inverse of $f(x) = \frac{x+1}{x-1}$ and determine its domain and range.

III Transcendental Functions, Matrices, and Systems of Equations

10 Transcendental Functions 203
- 10.1 Properties of Exponential and Logarithmic Functions
- 10.2 Solving Exponential and Logarithmic Equations
- 10.3 Trigonometric and Inverse Functions: Properties and Applications
- 10.4 Solved Exercises
- 10.5 Proposed Exercises

11 Matrices of Order 2 and 3 219
- 11.1 Matrix Operations (Addition, Subtraction, Multiplication)
- 11.2 Determinant Calculation and Adjoint of a Matrix
- 11.3 Inverse of a Matrix and Its Application in Systems of Equations
- 11.4 Ejercicios Resueltos
- 11.5 Solved Exercises
- 11.6 Proposed Exercises

12 Systems of Linear and Nonlinear Equations 243
- 12.1 Methods for Solving Linear Systems
- 12.2 Application of the Gauss-Jordan Method
- 12.3 Nonlinear Systems of Equations and Their Applications
- 12.4 Solved Exercises
- 12.5 Proposed Exercises

Índice Alfabético 263

10. Transcendental Functions

10.1 Properties of Exponential and Logarithmic Functions

10.1.1 Growth and Decay of Exponential Functions

Definition 10.1.1 An **exponential function** is a function of the form $f(x) = a^x$, where $a > 0$ and $a \neq 1$. This function exhibits exponential growth or decay depending on the value of a. If $a > 1$, the function $f(x)$ grows exponentially; if $0 < a < 1$, the function decays exponentially.

■ **Example 10.1** Consider the function $f(x) = 2^x$. Since $a = 2 > 1$, the function is increasing. For negative values of x, $f(x)$ approaches zero but remains positive. This rapid growth can be observed in values such as $f(1) = 2$, $f(2) = 4$, $f(3) = 8$, etc. ■

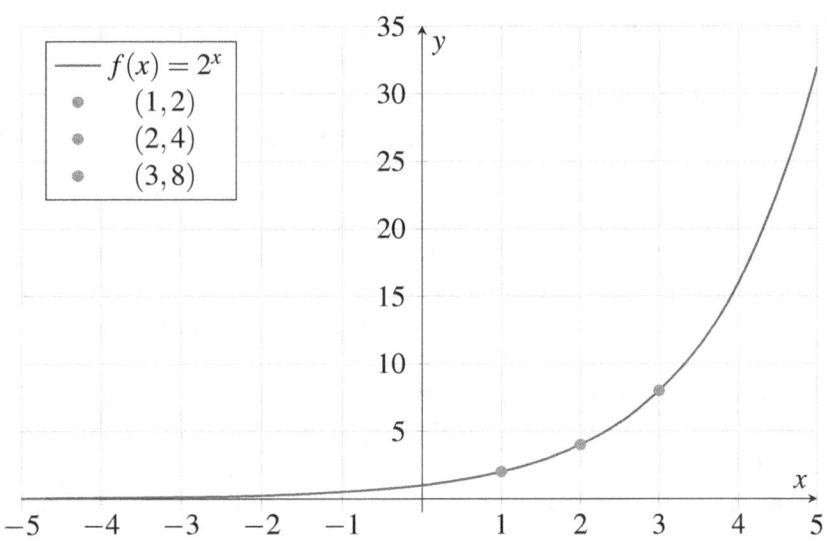

Figura 10.1.1: *Graph of the exponential function* $f(x) = 2^x$

Lema 10.1.1 The derivative of an exponential function of the form $f(x) = a^x$, where $a > 0$ and

$a \neq 1$, is given by $f'(x) = a^x \ln(a)$. This result indicates that the rate of change of the exponential function is proportional to its current value.

Demostración. Let $f(x) = a^x$, where $a > 0$ and $a \neq 1$. We aim to find the derivative of $f(x)$ with respect to x.
Using the definition of the derivative and the properties of the natural logarithm:

$$f(x) = a^x = e^{x \ln(a)}.$$

We apply the chain rule to differentiate $f(x) = e^{x \ln(a)}$:

$$f'(x) = \frac{d}{dx}\left(e^{x \ln(a)}\right) = e^{x \ln(a)} \cdot \ln(a).$$

Substituting $e^{x \ln(a)}$ with a^x, we get:

$$f'(x) = a^x \ln(a).$$

This shows that the derivative of $f(x) = a^x$ is $f'(x) = a^x \ln(a)$, indicating that the rate of change of $f(x)$ is proportional to its current value.

∎

Theorem 10.1.1 For any exponential function $f(x) = a^x$ with $a > 1$, the function is strictly increasing over the domain $\mathbb{R}$. This means that if $x_1 < x_2$, then $f(x_1) < f(x_2)$. Similarly, if $0 < a < 1$, the function is strictly decreasing.

Demostración. Consider the exponential function $f(x) = a^x$ and calculate its derivative to analyze its monotonicity.
1. The derivative of $f(x) = a^x$ is:

$$f'(x) = a^x \ln(a).$$

2. **Case $a > 1$**: - When $a > 1$, $\ln(a) > 0$. - Hence, $f'(x) = a^x \ln(a) > 0$ for all $x \in \mathbb{R}$. - This implies that $f(x)$ is strictly increasing over $\mathbb{R}$, and if $x_1 < x_2$, then $f(x_1) < f(x_2)$.
3. **Case $0 < a < 1$**: - When $0 < a < 1$, $\ln(a) < 0$. - In this case, $f'(x) = a^x \ln(a) < 0$ for all $x \in \mathbb{R}$. - This implies that $f(x)$ is strictly decreasing over $\mathbb{R}$, and if $x_1 < x_2$, then $f(x_1) > f(x_2)$.
Thus, if $a > 1$, the function $f(x) = a^x$ is strictly increasing, and if $0 < a < 1$, it is strictly decreasing.

∎

Corollary 10.1.2 Since the exponential function $f(x) = a^x$ with $a > 1$ is strictly increasing, we can conclude that the function is injective; that is, no two distinct values of x produce the same value of $f(x)$. This fact is important when considering the logarithmic inverse function.

Demostración. Let $f(x) = a^x$ with $a > 1$. As demonstrated in the previous theorem, $f(x)$ is strictly increasing over $\mathbb{R}$. This means that if $x_1 < x_2$, then $f(x_1) < f(x_2)$.
To verify injectivity, assume there exist $x_1, x_2 \in \mathbb{R}$ such that $f(x_1) = f(x_2)$. Since $f(x)$ is strictly increasing, this implies $x_1 = x_2$, as no two distinct values of x can produce the same $f(x)$.
Therefore, $f(x) = a^x$ is injective. This ensures that f has an inverse function, which is the corresponding logarithmic function.
□

10.1 Properties of Exponential and Logarithmic Functions

 Exponential functions are fundamental in modeling natural phenomena such as population growth, radioactive decay, and financial applications. Understanding their growth and decay behavior is essential for interpreting these phenomena in real-world contexts.

Exercise 10.1 Determine whether the function $f(x) = 0{,}5^x$ is increasing or decreasing. Explain your answer.

Exercise 10.2 Compute the derivative of the exponential function $f(x) = 3^x$ and determine the slope of the tangent at the point $x = 1$.

10.1.2 Properties of Logarithms

Definition 10.1.2 The **logarithm of a positive number** b to a base a (with $a > 0$ and $a \neq 1$) is the exponent to which a must be raised to obtain b. It is denoted by $\log_a(b)$ and is defined as:

$$\log_a(b) = x \quad \text{if and only if } a^x = b.$$

■ **Example 10.2** Consider the logarithm $\log_2(8)$. To determine its value, we find the exponent to which 2 must be raised to obtain 8. Since $2^3 = 8$, it follows that $\log_2(8) = 3$. ■

Lema 10.1.2 The logarithmic function $\log_a(x)$, with $a > 1$, is strictly increasing for $x > 0$. This implies that if $x_1 < x_2$, then $\log_a(x_1) < \log_a(x_2)$. This property is directly related to the behavior of the exponential function, as the logarithmic function is its inverse.

Demostración. Let $f(x) = \log_a(x)$, where $a > 1$. Recall that $f(x) = \log_a(x)$ is the inverse of the exponential function $g(x) = a^x$, which is strictly increasing for all $x \in \mathbb{R}$.
To prove that $f(x)$ is strictly increasing for $x > 0$, we compute its derivative. We know that:

$$f'(x) = \frac{d}{dx}\log_a(x) = \frac{1}{x\ln(a)}.$$

Since $a > 1$, we have $\ln(a) > 0$. Thus, for $x > 0$, the derivative $f'(x) = \frac{1}{x\ln(a)}$ is positive. This implies that $f(x) = \log_a(x)$ is strictly increasing on the interval $x > 0$.
Therefore, if $x_1 < x_2$, then $\log_a(x_1) < \log_a(x_2)$, confirming the increasing behavior of the logarithmic function. ∎

Theorem 10.1.3 The following are the **fundamental properties of logarithms**:
- **Product Rule**: $\log_a(xy) = \log_a(x) + \log_a(y)$.
- **Quotient Rule**: $\log_a\left(\frac{x}{y}\right) = \log_a(x) - \log_a(y)$.
- **Power Rule**: $\log_a(x^k) = k \cdot \log_a(x)$.

These properties apply whenever $x, y > 0$ and $a > 0, a \neq 1$.

Demostración. Let $f(x) = \log_a(x)$, where $a > 0$ and $a \neq 1$. We will prove each property using the definitions of logarithms and exponentials.
1. **Product Rule**: $\log_a(xy) = \log_a(x) + \log_a(y)$
- By the definition of logarithm, $\log_a(x) = p$ means $a^p = x$ and $\log_a(y) = q$ means $a^q = y$. - Thus, $xy = a^p \cdot a^q = a^{p+q}$. - By definition, $\log_a(xy) = p + q = \log_a(x) + \log_a(y)$.
2. **Quotient Rule**: $\log_a\left(\frac{x}{y}\right) = \log_a(x) - \log_a(y)$

- By definition, $\log_a(x) = p$ and $\log_a(y) = q$ imply $a^p = x$ and $a^q = y$. - Thus, $\frac{x}{y} = \frac{a^p}{a^q} = a^{p-q}$. - Therefore, $\log_a\left(\frac{x}{y}\right) = p - q = \log_a(x) - \log_a(y)$.

3. **Power Rule**: $\log_a(x^k) = k \cdot \log_a(x)$
- By definition, $\log_a(x) = p$ means $a^p = x$. - Thus, $x^k = (a^p)^k = a^{pk}$. - Therefore, $\log_a(x^k) = pk = k \cdot \log_a(x)$.

These properties hold for all $x, y > 0$ and $a > 0, a \neq 1$, completing the proof. ■

Corollary 10.1.4 The power property, $\log_a(x^k) = k \cdot \log_a(x)$, allows us to simplify expressions where a variable is raised to an exponent. This property is frequently used in solving logarithmic and exponential equations.

(R) These properties are fundamental for simplifying and solving logarithmic equations. Additionally, they play a significant role in calculus and the analysis of phenomena involving exponential growth and logarithmic scales.

Exercise 10.3 Simplify the expression $\log_3(27) + \log_3(9)$ using the properties of logarithms. ■

Exercise 10.4 Solve the equation $\log_5(x) + \log_5(4) = 2$. Determine the value of x. ■

10.1.3 Graphs of Exponential and Logarithmic Functions

Definition 10.1.3 The **exponential function** with base $a > 0$, $a \neq 1$, is defined as $f(x) = a^x$. The graph of an exponential function exhibits characteristic growth or decay depending on the value of the base a. The **logarithmic function**, denoted as $g(x) = \log_a(x)$, is the inverse of the exponential function $f(x) = a^x$.

■ **Example 10.3** Consider the exponential function $f(x) = 2^x$ and the logarithmic function $g(x) = \log_2(x)$. These functions are inverses of each other, which is reflected in the symmetry of their graphs with respect to the line $y = x$. ■

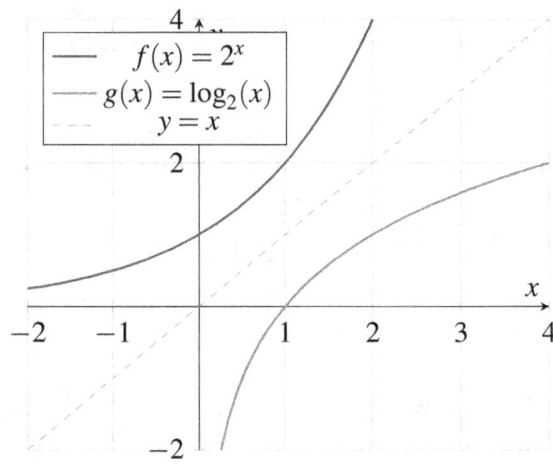

Lema 10.1.3 For the exponential function $f(x) = a^x$ with $a > 1$, the graph is always increasing, and for $0 < a < 1$, the graph is decreasing. This implies that exponential functions are injective, and therefore have well-defined inverses.

10.1 Properties of Exponential and Logarithmic Functions

Demostración. Let $f(x) = a^x$, where $a > 0$ and $a \neq 1$. We compute the derivative of $f(x)$ to determine its monotonicity:

$$f'(x) = a^x \ln(a).$$

1. **Case $a > 1$**: - When $a > 1$, we have $\ln(a) > 0$. - Thus, $f'(x) = a^x \ln(a) > 0$ for all $x \in \mathbb{R}$, implying that $f(x)$ is strictly increasing over its domain.
2. **Case $0 < a < 1$**: - When $0 < a < 1$, we have $\ln(a) < 0$. - Thus, $f'(x) = a^x \ln(a) < 0$ for all $x \in \mathbb{R}$, implying that $f(x)$ is strictly decreasing over its domain.

Since $f(x)$ is strictly increasing when $a > 1$ and strictly decreasing when $0 < a < 1$, the function $f(x) = a^x$ is injective in both cases. This ensures that exponential functions have well-defined inverses. ∎

> **Theorem 10.1.5** The graph of the **logarithmic function** $g(x) = \log_a(x)$ has the following properties:
> - Its domain is $(0, \infty)$, and its range is $(-\infty, \infty)$.
> - It passes through the point $(1, 0)$, as $\log_a(1) = 0$.
> - If $a > 1$, the function is increasing; if $0 < a < 1$, the function is decreasing.
>
> These properties follow from its inverse relationship with the exponential function.

Demostración. Consider the logarithmic function $g(x) = \log_a(x)$, where $a > 0$ and $a \neq 1$. We will demonstrate each stated property.

1. **Domain and Range**: The function $g(x) = \log_a(x)$ is the inverse of the exponential function $f(x) = a^x$, whose graph has a domain of $\mathbb{R}$ and a range of $(0, \infty)$. - Since $g(x)$ is the inverse of $f(x)$, the domain of $g(x)$ is the range of $f(x)$, which is $(0, \infty)$. - The range of $g(x)$ is the domain of $f(x)$, which is $\mathbb{R}$. - Therefore, the domain of $g(x)$ is $(0, \infty)$, and its range is $(-\infty, \infty)$.
2. **Passes through the point $(1, 0)$**: - Since $a^0 = 1$, it follows that $\log_a(1) = 0$. - Thus, the graph of $g(x)$ passes through the point $(1, 0)$.
3. **Increasing or Decreasing Behavior of $g(x)$**: - The derivative of $g(x) = \log_a(x)$ is $g'(x) = \frac{1}{x \ln(a)}$. - If $a > 1$, then $\ln(a) > 0$, and thus $g'(x) > 0$ for $x > 0$, implying that $g(x)$ is increasing. - If $0 < a < 1$, then $\ln(a) < 0$, and thus $g'(x) < 0$ for $x > 0$, implying that $g(x)$ is decreasing.

These properties confirm the behavior of the graph of $g(x) = \log_a(x)$, derived from its inverse relationship with the exponential function $f(x) = a^x$. ∎

> **Corollary 10.1.6** Since exponential and logarithmic functions are inverses of each other, the graph of a logarithmic function can be obtained by reflecting the graph of the corresponding exponential function about the line $y = x$. This symmetry helps to understand the relationship between their growth and decay rates.

> (R) The relationship between the graphs of exponential and logarithmic functions provides a deeper understanding of how these functions behave in different contexts, such as population growth or the decibel scale for measuring sound.

> **Exercise 10.5** Draw the graph of the function $f(x) = 3^x$ and determine how the graph changes if the base is reduced to $a = 1/3$.

> **Exercise 10.6** Determine the domain and range of the function $g(x) = \log_5(x-2)$. How would this change be reflected in the graph of the standard logarithmic function $g(x) = \log_5(x)$?

These exercises will help students connect the theoretical properties of exponential and logarithmic functions with their graphical representations, reinforcing both visual and analytical understanding of the concepts involved.

10.2 Solving Exponential and Logarithmic Equations

10.2.1 Basic Exponential Equations

> **Definition 10.2.1** An **exponential equation** is one where the unknown appears in the exponent, generally in the form $a^x = b$, where $a > 0$ and $a \neq 1$. Solving an exponential equation involves finding the value of x that satisfies the equation.

■ **Example 10.4** Let us solve the exponential equation $3^{x+1} = 27$. We know that 27 can be expressed as a power of 3, specifically $27 = 3^3$. Thus:

$$3^{x+1} = 3^3$$

By comparing the exponents, we have $x + 1 = 3$, which implies $x = 2$. ■

Lema 10.2.1 If $a^x = a^y$ for $a > 0$ and $a \neq 1$, then necessarily $x = y$. This property is known as the **uniqueness of equality of exponents** and is fundamental for solving exponential equations with the same base.

Demostración. Let $a > 0$ and $a \neq 1$. Suppose $a^x = a^y$. We aim to prove that this equality implies $x = y$.
Consider the function $f(x) = a^x$, which is strictly monotonic (increasing if $a > 1$ and decreasing if $0 < a < 1$). Due to this strict monotonicity, $f(x)$ is injective, meaning that if $f(x) = f(y)$, then $x = y$.
Given $a^x = a^y$, which implies $f(x) = f(y)$, we conclude $x = y$. ∎

> **Theorem 10.2.1** Any exponential equation of the form $a^{f(x)} = b$ can be solved by taking logarithms on both sides of the equation, provided $a > 0$, $a \neq 1$, and $b > 0$. This is based on the logarithmic property that allows exponents to be "brought down," facilitating the solution.

Demostración. Consider the exponential equation $a^{f(x)} = b$, where $a > 0$, $a \neq 1$, and $b > 0$. To solve this equation, take logarithms on both sides:

$$\log_a(a^{f(x)}) = \log_a(b).$$

Using the logarithmic property that allows the exponent to be brought down, we obtain:

$$f(x) \cdot \log_a(a) = \log_a(b).$$

Since $\log_a(a) = 1$, this simplifies to:

$$f(x) = \log_a(b).$$

Thus, we solve for $f(x)$ by expressing the right-hand side in terms of the logarithm in base a. This technique is valid as long as $a > 0$, $a \neq 1$, and $b > 0$, enabling exponential equations to be solved using logarithms. ∎

10.2 Solving Exponential and Logarithmic Equations

Corollary 10.2.2 Since the logarithmic function $f(x) = \log_a(x)$ is the inverse of the exponential function $g(x) = a^x$, exponential equations can be solved by rewriting them in terms of logarithms. This provides an additional tool for handling more complex equations.

Demostración. Given that the logarithmic function $f(x) = \log_a(x)$ is the inverse of the exponential function $g(x) = a^x$, the following holds:

$$f(g(x)) = \log_a(a^x) = x \quad \text{and} \quad g(f(x)) = a^{\log_a(x)} = x.$$

This implies that, to solve an exponential equation of the form $a^{f(x)} = b$, we can apply the logarithm in base a to both sides:

$$\log_a(a^{f(x)}) = \log_a(b).$$

Applying the logarithm yields:

$$f(x) = \log_a(b).$$

Thus, exponential equations can be rewritten in terms of logarithms, simplifying their resolution. This technique is made possible by the inverse relationship between a^x and $\log_a(x)$, providing a useful tool for solving exponential equations. ∎

> (R) It is important to verify that the solutions obtained for exponential equations are valid within the domain of the original equation.

Exercise 10.7 Solve the exponential equation $2^{x-1} = 16$. Use the uniqueness property of equality of exponents to find the value of x.

Exercise 10.8 Find the solution to the equation $5^{2x} = 125$ by taking logarithms on both sides and simplifying to isolate x.

These exercises and results illustrate how the properties of exponential equations are used in various contexts, creating a natural bridge to solving logarithmic equations and studying transcendental functions.

10.2.2 Logarithmic Equations and Their Properties

Definition 10.2.2 A **logarithmic equation** is one that involves logarithms in its expressions, generally of the form $\log_a(f(x)) = b$, where $a > 0$ and $a \neq 1$. Solving a logarithmic equation involves finding the value of the variable that satisfies the equation.

■ **Example 10.5** Consider the logarithmic equation $\log_3(x+2) = 2$. To solve it, we first rewrite the equation in exponential form:

$$x + 2 = 3^2$$

This gives $x + 2 = 9$, from which we obtain $x = 7$. ■

Lema 10.2.2 If $\log_a(f(x)) = \log_a(g(x))$ for $a > 0$ and $a \neq 1$, then $f(x) = g(x)$. This property, known as the **uniqueness of equality of logarithms**, is fundamental for solving logarithmic equations with the same base.

Demostración. Let $a > 0$ and $a \neq 1$. Suppose $\log_a(f(x)) = \log_a(g(x))$. We aim to prove that this equality implies $f(x) = g(x)$.

The logarithmic function $\log_a(x)$ is strictly monotonic (increasing if $a > 1$ and decreasing if $0 < a < 1$). Due to this strict monotonicity, $\log_a(x)$ is injective, meaning that if $\log_a(f(x)) = \log_a(g(x))$, then necessarily $f(x) = g(x)$.

Thus, we conclude that $f(x) = g(x)$, as required. ∎

> **Theorem 10.2.3** To solve a logarithmic equation of the form $\log_a(f(x)) = b$, it can be rewritten in exponential form: $f(x) = a^b$. This technique allows us to convert a logarithmic equation into an exponential one, facilitating its resolution.

Demostración. Consider the logarithmic equation $\log_a(f(x)) = b$, where $a > 0$ and $a \neq 1$. We aim to solve this equation for $f(x)$.

By the definition of a logarithm, $\log_a(f(x)) = b$ means that the exponent to which a must be raised to obtain $f(x)$ is b. This can be rewritten in exponential form as:

$$f(x) = a^b.$$

By expressing the equation in this exponential form, we transform the logarithmic equation into an exponential one, making it easier to solve for $f(x)$.

Thus, the logarithmic equation $\log_a(f(x)) = b$ is equivalent to the exponential equation $f(x) = a^b$, allowing us to solve for $f(x)$ directly. ∎

> **Corollary 10.2.4** Since logarithmic and exponential functions are inverses of each other, any logarithmic equation can be solved by applying properties of exponentiation, as long as the domain restrictions of the logarithmic function are respected.

(R) It is important to remember that the argument of a logarithm must always be positive, i.e., $f(x) > 0$. This restricts the domain of solutions for logarithmic equations.

> **Exercise 10.9** Solve the equation $\log_2(x-1) = 3$ and verify that the solution lies within the allowed domain.

> **Exercise 10.10** Find all values of x that satisfy the equation $\log_5(x) + \log_5(2) = 1$. Use logarithmic properties to simplify and solve.

These results demonstrate how logarithmic properties allow for simplifying and solving complex equations. Additionally, the connection between exponential and logarithmic equations is key to understanding the resolution of both classes of equations.

10.2.3 Applications in Growth and Decay Problems

> **Definition 10.2.3 Exponential growth** occurs when a quantity increases at a rate proportional to its current size, while **exponential decay** describes a proportional decrease over time. Both scenarios are modeled with functions of the form $y = Ae^{kt}$, where A is the initial amount, k is the growth rate (positive for growth and negative for decay), and t is time.

■ **Example 10.6** A typical example of exponential growth is bacterial population growth. Suppose a bacterial population doubles every hour. If we start with 100 bacteria, the population after t hours can be modeled by the equation:

$$P(t) = 100 \times 2^t$$

This model follows the structure of exponential growth, where $A = 100$ and the base 2 represents growth at a doubling rate. ■

Lema 10.2.3 If a function $f(t)$ describes the exponential growth or decay of a quantity, then its derivative $f'(t)$ also has the same exponential form, with a constant that depends on the growth or decay rate.

Theorem 10.2.5 The **Exponential Growth and Decay Theorem** states that if $y(t) = Ae^{kt}$, then the quantity will increase (if $k > 0$) or decrease (if $k < 0$) exponentially over time. This function appropriately models processes such as radioactive decay, population growth, and Newton's law of cooling.

Corollary 10.2.6 For any quantity that decays at a rate proportional to its current value, its decay time can be determined using the decay constant k. For instance, the half-life is defined as $T_{1/2} = \frac{\ln(2)}{k}$, which is particularly relevant in radioactive decay.

(R) Exponential growth and decay are applied in various fields such as biology, economics, and physics. A common feature of these applications is the multiplicative nature of change, where the present quantity directly influences the rate of change.

Exercise 10.11 A certain radioactive substance has a half-life of 10 years. If there are initially 100 grams of the substance, how much will remain after 30 years?

Exercise 10.12 The population of a city grows at a rate proportional to the number of inhabitants. If the initial population is 5,000 people and grows at a rate of 3% per year, find the population after 10 years.

These results demonstrate how exponential functions can model the behavior of real systems where change depends proportionally on the system's current state. The exercises provide practical examples of how to apply these concepts to real-world problems.

10.3 Trigonometric and Inverse Functions: Properties and Applications

10.3.1 Definition of Trigonometric Functions

Definition 10.3.1 Trigonometric functions relate the angles of a right triangle to the lengths of its sides. The basic trigonometric functions are sine, cosine, and tangent, defined as follows for an angle θ in a right triangle: - $\sin(\theta) = \frac{\text{opposite side}}{\text{hypotenuse}}$ - $\cos(\theta) = \frac{\text{adjacent side}}{\text{hypotenuse}}$ - $\tan(\theta) = \frac{\text{opposite side}}{\text{adjacent side}}$

■ **Example 10.7** Consider a right triangle with a 30° angle, a hypotenuse of length 10, and an opposite side of length 5. We can calculate the sine, cosine, and tangent of 30° as follows:

$$\sin(30°) = \frac{5}{10} = 0.5, \quad \cos(30°) = \frac{\sqrt{75}}{10}, \quad \tan(30°) = \frac{5}{\sqrt{75}}$$

This example demonstrates how trigonometric functions apply to specific triangles to determine relationships between their sides. ∎

Lema 10.3.1 The value of the trigonometric function $\sin(\theta)$ is always a number between -1 and 1 for any angle θ. The same applies to the function $\cos(\theta)$.

> **Theorem 10.3.1** For any angle θ, the following trigonometric identity holds:
>
> $$\sin^2(\theta) + \cos^2(\theta) = 1$$
>
> This is known as the **Pythagorean identity**, and it is valid for all real values of θ. The identity is useful for simplifying many trigonometric expressions and proving other properties.

Demostración. Consider an angle θ in a unit circle, where the radius of the circle is 1. In the unit circle, any point on the circumference corresponding to the angle θ has coordinates $(\cos(\theta), \sin(\theta))$. By the Pythagorean Theorem, the distance from the origin $(0,0)$ to the point $(\cos(\theta), \sin(\theta))$ equals 1, so:

$$\cos^2(\theta) + \sin^2(\theta) = 1^2 = 1.$$

Thus, we obtain the identity:

$$\sin^2(\theta) + \cos^2(\theta) = 1.$$

This identity is valid for all real values of θ and is derived directly from the properties of the unit circle. ∎

> **Corollary 10.3.2** Given that $\sin^2(\theta) + \cos^2(\theta) = 1$, we can derive that:
>
> $$\tan(\theta) = \frac{\sin(\theta)}{\cos(\theta)}$$
>
> provided $\cos(\theta) \neq 0$. This relationship is fundamental to understanding the connection between the three basic trigonometric functions.

Demostración. From the Pythagorean identity, we know:

$$\sin^2(\theta) + \cos^2(\theta) = 1.$$

Divide both sides of this equation by $\cos^2(\theta)$, assuming $\cos(\theta) \neq 0$:

$$\frac{\sin^2(\theta)}{\cos^2(\theta)} + \frac{\cos^2(\theta)}{\cos^2(\theta)} = \frac{1}{\cos^2(\theta)}.$$

Simplifying, we obtain:

$$\tan^2(\theta) + 1 = \sec^2(\theta),$$

where $\tan(\theta) = \frac{\sin(\theta)}{\cos(\theta)}$ and $\sec(\theta) = \frac{1}{\cos(\theta)}$. This shows that the tangent can be expressed as the ratio of sine to cosine, establishing a relationship between these trigonometric functions. ∎

> (R) Trigonometric functions are cyclical and have applications across a wide variety of fields, from physics (waves) to engineering (periodic motion).

10.3 Trigonometric and Inverse Functions: Properties and Applications

Exercise 10.13 Prove that $\sin(2\theta) = 2\sin(\theta)\cos(\theta)$ using the Pythagorean identity and the definitions of trigonometric functions.

Exercise 10.14 Calculate the exact value of $\tan(45°)$ and explain why this value remains constant regardless of the size of the considered triangle.

The theorem on the Pythagorean identity naturally leads to exploring how other trigonometric identities, such as the double-angle formula, are derived. The exercises help to understand how these properties apply in specific scenarios.

10.3.2 Fundamental Relationships Between Trigonometric Functions

Definition 10.3.2 The **fundamental relationships between trigonometric functions** connect sine, cosine, tangent, cotangent, secant, and cosecant. Some of the most well-known relationships are:

$$\sin^2(\theta) + \cos^2(\theta) = 1$$

$$\tan(\theta) = \frac{\sin(\theta)}{\cos(\theta)}, \quad \cot(\theta) = \frac{\cos(\theta)}{\sin(\theta)}$$

■ **Example 10.8** Consider the angle $\theta = 45°$. Using the fundamental relationships:

$$\sin^2(45°) + \cos^2(45°) = \left(\frac{\sqrt{2}}{2}\right)^2 + \left(\frac{\sqrt{2}}{2}\right)^2 = \frac{1}{2} + \frac{1}{2} = 1$$

This example verifies that the Pythagorean identity holds for the given angle. ■

Lema 10.3.2 For any angle θ, the secant function and the cosine function are related as follows:

$$\sec(\theta) = \frac{1}{\cos(\theta)}$$

provided $\cos(\theta) \neq 0$. This relationship is essential for defining the secant function in terms of the basic trigonometric functions.

Demostración. By definition, the secant function of an angle θ is defined as the reciprocal of the cosine of θ, that is:

$$\sec(\theta) = \frac{1}{\cos(\theta)}.$$

This relationship is valid as long as $\cos(\theta) \neq 0$, because in this case, the reciprocal of $\cos(\theta)$ is well-defined.

Thus, the secant function is expressed in terms of the cosine function as $\sec(\theta) = \frac{1}{\cos(\theta)}$, establishing a direct connection between these two trigonometric functions. ■

Theorem 10.3.3 For any angle θ, the following identities hold:

$$1 + \tan^2(\theta) = \sec^2(\theta)$$

$$1 + \cot^2(\theta) = \csc^2(\theta)$$

These identities are directly derived from the Pythagorean identity and are useful for simplifying expressions involving tangent and cotangent.

Demostración. To prove these identities, we start from the basic Pythagorean identity:

$$\sin^2(\theta) + \cos^2(\theta) = 1.$$

1. **Proof of $1 + \tan^2(\theta) = \sec^2(\theta)$**

Divide both sides of the Pythagorean identity by $\cos^2(\theta)$, assuming $\cos(\theta) \neq 0$:

$$\frac{\sin^2(\theta)}{\cos^2(\theta)} + \frac{\cos^2(\theta)}{\cos^2(\theta)} = \frac{1}{\cos^2(\theta)}.$$

Simplifying, we obtain:

$$\tan^2(\theta) + 1 = \sec^2(\theta).$$

2. **Proof of $1 + \cot^2(\theta) = \csc^2(\theta)$**

Divide both sides of the Pythagorean identity by $\sin^2(\theta)$, assuming $\sin(\theta) \neq 0$:

$$\frac{\sin^2(\theta)}{\sin^2(\theta)} + \frac{\cos^2(\theta)}{\sin^2(\theta)} = \frac{1}{\sin^2(\theta)}.$$

Simplifying, we obtain:

$$1 + \cot^2(\theta) = \csc^2(\theta).$$

These identities are directly derived from the Pythagorean identity and are valid as long as $\cos(\theta) \neq 0$ and $\sin(\theta) \neq 0$. ∎

Corollary 10.3.4 Given that $1 + \tan^2(\theta) = \sec^2(\theta)$, we can infer that for $\theta = 45°$:

$$1 + \tan^2(45°) = \sec^2(45°) \Rightarrow 1 + 1 = \sec^2(45°) = 2$$

This result confirms the validity of the identity for a specific value of θ.

R The fundamental relationships between trigonometric functions are key to solving trigonometric equations and simplifying complex expressions in mathematical calculations and physics problems.

Exercise 10.15 Prove that $1 + \cot^2(\theta) = \csc^2(\theta)$ using the Pythagorean identity and the definitions of trigonometric functions.

10.3 Trigonometric and Inverse Functions: Properties and Applications

Exercise 10.16 Calculate the exact value of $\sec(60°)$ and prove that it satisfies the relationship $\sec(\theta) = \frac{1}{\cos(\theta)}$.

The trigonometric identities presented in this chapter allow us to establish connections between the functions and solve problems more efficiently, as demonstrated in the examples and proposed exercises.

10.3.3 Inverse Trigonometric Functions and Their Applications

Definition 10.3.3 **Inverse trigonometric functions** allow us to determine the angle whose trigonometric function value corresponds to a known value. The most common inverse trigonometric functions are:
- $\arcsin(x)$ or $\sin^{-1}(x)$: Inverse of the sine function. - $\arccos(x)$ or $\cos^{-1}(x)$: Inverse of the cosine function. - $\arctan(x)$ or $\tan^{-1}(x)$: Inverse of the tangent function.
These functions are defined over specific intervals where the original functions are bijective, ensuring the existence of a unique inverse.

■ **Example 10.9** Calculate the value of θ if $\cos(\theta) = \frac{1}{2}$ and $0 \leq \theta \leq \pi$.
Using the inverse cosine function, we have:

$$\theta = \arccos\left(\frac{1}{2}\right) = \frac{\pi}{3}$$

This indicates that the angle in the first quadrant whose cosine is $\frac{1}{2}$ is $\frac{\pi}{3}$. ■

Lema 10.3.3 For any value x within the domain of $\arcsin(x)$, the following property holds:

$$\sin(\arcsin(x)) = x, \quad \text{for } -1 \leq x \leq 1$$

This ensures that the composition of the sine function with its inverse returns the original value of x, provided that x is within the domain of $\arcsin(x)$.

Theorem 10.3.5 If $\theta = \arcsin(x)$, then:

$$\cos(\theta) = \sqrt{1-x^2}, \quad \text{for } -1 \leq x \leq 1$$

This theorem allows us to compute the cosine of an angle given its sine value, using the Pythagorean theorem. This property is fundamental in solving trigonometric problems.

Corollary 10.3.6 Given that $\tan(\arctan(x)) = x$ for any real number x, we can conclude that the functions tan and arctan cancel each other within their domains, similar to what is observed for sin and arcsin.

(R) It is important to remember that inverse trigonometric functions are defined over specific intervals to ensure their uniqueness. For example, $\arcsin(x)$ is defined on the interval $[-\frac{\pi}{2}, \frac{\pi}{2}]$ to guarantee that every input value has a unique output.

Exercise 10.17 Find the value of θ if $\tan(\theta) = 1$ and $-\frac{\pi}{2} < \theta < \frac{\pi}{2}$.

Exercise 10.18 Solve the equation $\arccos(x) = \frac{\pi}{4}$. Determine the value of x and verify that it lies within the appropriate domain.

In this section, we have introduced the definitions and applications of inverse trigonometric functions. In particular, we highlighted their fundamental properties and the relationships between trigonometric functions and their inverses. These properties simplify solving trigonometric equations and facilitate understanding the geometry and analysis of angles in diverse contexts, from physics to engineering.

10.4 Solved Exercises

Exercise 10.19 Determine the domain of the exponential function $f(x) = e^{2x-5}$.

Demostración. The exponential function is defined for all real numbers. Therefore, the domain is $\mathbb{R}$. ∎

Exercise 10.20 Calculate the value of $\ln(20)$ rounded to two decimal places.

Demostración. Using a calculator, we find $\ln(20) \approx 2{,}9957$. Rounded to two decimal places, the result is $2{,}99$. ∎

Exercise 10.21 Solve the exponential equation $3^{x+1} = 27$.

Demostración. We can write 27 as 3^3. Thus, the equation becomes $3^{x+1} = 3^3$. Equating the exponents, we obtain $x + 1 = 3$, which implies $x = 2$. ∎

Exercise 10.22 Calculate the derivative of the logarithmic function $f(x) = \ln(2x+1)$.

Demostración. Using the differentiation rule for logarithms, we have:

$$f'(x) = \frac{1}{2x+1} \cdot 2 = \frac{2}{2x+1}.$$

∎

Exercise 10.23 Find the value of x in the equation $e^{x^2} = 5e$.

Demostración. Dividing both sides by e, we get $e^{x^2-1} = 5$. Taking the natural logarithm, we obtain $x^2 = 1 + \ln(5)$, so $x = \pm\sqrt{1 + \ln(5)}$. ∎

10.5 Proposed Exercises

10.5.1 Properties of Exponential and Logarithmic Functions

Exercise 10.24 Prove that the function $f(x) = e^x$ is always increasing.

10.5 Proposed Exercises

Exercise 10.25 Find the domain and range of the function $f(x) = \ln(x-3)$.

Exercise 10.26 Calculate the limit $\lim_{x \to \infty} e^{-x}$ and describe the behavior of the function.

Exercise 10.27 Determine whether the function $g(x) = -e^{-x}$ is increasing or decreasing.

Exercise 10.28 Solve the equation $e^{2x} = 7$ and express the result for x.

10.5.2 Solving Exponential and Logarithmic Equations

Exercise 10.29 Solve the equation $\ln(x) + \ln(x-1) = 1$.

Exercise 10.30 Calculate x for the equation $e^{3x} = 10$.

Exercise 10.31 Simplify and solve the equation $2\log(x) = \log(16)$.

Exercise 10.32 Find the solution to the equation $e^x = 5x$.

Exercise 10.33 Solve the equation $\ln(x^2) - \ln(x) = 3$.

10.5.3 Trigonometric and Inverse Functions: Properties and Applications

Exercise 10.34 Calculate the value of $\sin^{-1}\left(\frac{1}{2}\right)$.

Exercise 10.35 Find the domain of the function $f(x) = \cos^{-1}(x)$.

Exercise 10.36 Determine $\tan(\tan^{-1}(x))$ and explain the result.

Exercise 10.37 Solve the equation $\sin^{-1}(x) = \frac{\pi}{6}$.

Exercise 10.38 If $f(x) = \sin(x)$ and $g(x) = \cos^{-1}(x)$, calculate $(f \circ g)(x)$.

11. Matrices of Order 2 and 3

11.1 Matrix Operations (Addition, Subtraction, Multiplication)

11.1.1 Matrix Addition and Subtraction

Definition 11.1.1 The **addition of matrices** involves adding element by element of two matrices of the same order. If we have two matrices $A = [a_{ij}]$ and $B = [b_{ij}]$ of order $m \times n$, their sum is a matrix $C = [c_{ij}]$, where each element is defined as:

$$c_{ij} = a_{ij} + b_{ij}$$

Similarly, the **subtraction of matrices** is performed by subtracting the corresponding elements of matrices of the same order. The resulting matrix is $D = [d_{ij}]$, where:

$$d_{ij} = a_{ij} - b_{ij}$$

It is important to note that both matrices must have the same order to be added or subtracted.

■ **Example 11.1** Consider the matrices:

$$A = \begin{bmatrix} 2 & 3 \\ 4 & 5 \end{bmatrix}, \quad B = \begin{bmatrix} 1 & 0 \\ -1 & 3 \end{bmatrix}$$

The sum of A and B is:

$$A + B = \begin{bmatrix} 2+1 & 3+0 \\ 4+(-1) & 5+3 \end{bmatrix} = \begin{bmatrix} 3 & 3 \\ 3 & 8 \end{bmatrix}$$

And the subtraction of A and B is:

$$A - B = \begin{bmatrix} 2-1 & 3-0 \\ 4-(-1) & 5-3 \end{bmatrix} = \begin{bmatrix} 1 & 3 \\ 5 & 2 \end{bmatrix}$$

∎

Lema 11.1.1 For any pair of matrices A and B of the same order, the addition operation is commutative, i.e.:

$$A + B = B + A$$

This is because the addition of the elements of the matrices is performed element by element, and the addition of real numbers is commutative.

Demostración. Let $A = [a_{ij}]$ and $B = [b_{ij}]$ be two matrices of the same order $m \times n$. The sum $A + B$ is defined as the matrix $C = [c_{ij}]$, where each element c_{ij} is obtained by adding the corresponding elements of A and B:

$$c_{ij} = a_{ij} + b_{ij}.$$

Similarly, the sum $B + A$ is the matrix $D = [d_{ij}]$, where:

$$d_{ij} = b_{ij} + a_{ij}.$$

Since the addition of real numbers is commutative, we have:

$$a_{ij} + b_{ij} = b_{ij} + a_{ij}.$$

Thus, $c_{ij} = d_{ij}$ for all i and j, which implies $A + B = B + A$. ∎

Theorem 11.1.1 Let A, B, C be matrices of the same order. Then, matrix addition is associative, i.e.:

$$A + (B + C) = (A + B) + C$$

The associative property of matrix addition ensures that the grouping of matrices does not affect the final result of the addition operation.

Demostración. Let $A = [a_{ij}]$, $B = [b_{ij}]$, and $C = [c_{ij}]$ be matrices of the same order $m \times n$. We want to prove that $A + (B + C) = (A + B) + C$.
The sum $B + C$ is the matrix $D = [d_{ij}]$, where each element d_{ij} is defined as:

$$d_{ij} = b_{ij} + c_{ij}.$$

Then, $A + (B + C)$ is the matrix $E = [e_{ij}]$, where:

$$e_{ij} = a_{ij} + d_{ij} = a_{ij} + (b_{ij} + c_{ij}).$$

On the other hand, the sum $A + B$ is the matrix $F = [f_{ij}]$, where:

11.1 Matrix Operations (Addition, Subtraction, Multiplication)

$$f_{ij} = a_{ij} + b_{ij}.$$

Next, $(A+B)+C$ is the matrix $G = [g_{ij}]$, where:

$$g_{ij} = f_{ij} + c_{ij} = (a_{ij} + b_{ij}) + c_{ij}.$$

Since the addition of real numbers is associative, we have:

$$a_{ij} + (b_{ij} + c_{ij}) = (a_{ij} + b_{ij}) + c_{ij}.$$

Thus, $e_{ij} = g_{ij}$ for all i and j, which implies $A + (B+C) = (A+B) + C$. ∎

Corollary 11.1.2 Given any matrix A, there exists a **zero matrix** O of the same order such that:

$$A + O = A$$

The zero matrix has all its elements equal to zero and acts as the additive identity for matrix addition.

Demostración. Let $A = [a_{ij}]$ be a matrix of order $m \times n$. Define the zero matrix $O = [o_{ij}]$ of the same order, where each element $o_{ij} = 0$.
The sum $A + O$ is the matrix $C = [c_{ij}]$, where each element c_{ij} is calculated as:

$$c_{ij} = a_{ij} + o_{ij} = a_{ij} + 0 = a_{ij}.$$

Thus, each element of C is equal to the corresponding element of A, which implies $C = A$. This demonstrates that $A + O = A$, confirming that the zero matrix O acts as the additive identity for matrix addition. ∎

(R) Matrix addition and subtraction are only defined if both matrices have the same order. Otherwise, the operations cannot be performed. This restriction is important to ensure that each element has a corresponding element in the other matrix.

Exercise 11.1 Calculate the addition and subtraction of the following matrices:

$$A = \begin{bmatrix} 1 & -2 & 3 \\ 0 & 4 & 5 \end{bmatrix}, \quad B = \begin{bmatrix} -1 & 3 & 2 \\ 1 & 0 & -3 \end{bmatrix}$$

Exercise 11.2 Verify whether the commutative property of addition holds for the matrices:

$$A = \begin{bmatrix} 5 & 2 \\ -1 & 3 \end{bmatrix}, \quad B = \begin{bmatrix} 0 & 4 \\ 3 & 1 \end{bmatrix}$$

■ **Example 11.2** The following example illustrates the addition of two matrices:

$$A = \begin{bmatrix} 2 & 3 \\ 4 & 5 \end{bmatrix}, \quad B = \begin{bmatrix} 1 & 1 \\ 0 & -2 \end{bmatrix}$$

The result of the sum $A + B$ is:

$$\begin{bmatrix} 3 & 4 \\ 4 & 3 \end{bmatrix}$$

11.1.2 Matrix Multiplication and Properties

Definition 11.1.2 **Matrix multiplication** combines two matrices, A of order $m \times n$ and B of order $n \times p$, to produce a new matrix C of order $m \times p$. Each element of C, denoted as c_{ij}, is calculated as the sum of the products of the elements in row i of A and the corresponding elements in column j of B:

$$c_{ij} = \sum_{k=1}^{n} a_{ik} b_{kj}$$

It is important to note that matrix multiplication is not commutative, i.e., $AB \neq BA$ in general.

■ **Example 11.3** Consider the matrices:

$$A = \begin{bmatrix} 1 & 2 \\ 3 & 4 \end{bmatrix}, \quad B = \begin{bmatrix} 2 & 0 \\ 1 & 3 \end{bmatrix}$$

The product of A and B is:

$$AB = \begin{bmatrix} (1 \cdot 2 + 2 \cdot 1) & (1 \cdot 0 + 2 \cdot 3) \\ (3 \cdot 2 + 4 \cdot 1) & (3 \cdot 0 + 4 \cdot 3) \end{bmatrix} = \begin{bmatrix} 4 & 6 \\ 10 & 12 \end{bmatrix}$$

Lema 11.1.2 Matrix multiplication is **associative**, meaning that for any three matrices A, B, and C, the following holds:

$$A(BC) = (AB)C$$

This property guarantees that the grouping of matrices does not affect the result of the product.

Demostración. Let A be a matrix of dimension $m \times n$, B of dimension $n \times p$, and C of dimension $p \times q$. We want to prove that:

$$A(BC) = (AB)C.$$

For any entry (i, j), we have:

$$(A(BC))_{ij} = \sum_{k=1}^{n} A_{ik} \left(\sum_{l=1}^{p} B_{kl} C_{lj} \right).$$

Using the commutative property of summation, this equals:

$$(A(BC))_{ij} = \sum_{l=1}^{p} \left(\sum_{k=1}^{n} A_{ik} B_{kl} \right) C_{lj} = ((AB)C)_{ij}.$$

Thus, $A(BC) = (AB)C$, demonstrating associativity.

11.1 Matrix Operations (Addition, Subtraction, Multiplication)

> **Theorem 11.1.3** Let A be a matrix of order $m \times n$ and I_n the identity matrix of order $n \times n$. Then:
>
> $$AI_n = A$$
>
> and if I_m is the identity matrix of order $m \times m$:
>
> $$I_m A = A$$
>
> The identity matrix acts as the **neutral element** for matrix multiplication.

Demostración. Let $A = [a_{ij}]$ be a matrix of order $m \times n$, and let I_n be the identity matrix of order $n \times n$, where $I_n = [\delta_{jk}]$, with $\delta_{jk} = 1$ if $j = k$ and $\delta_{jk} = 0$ otherwise.
To prove $AI_n = A$, consider the matrix product. The element at position (i, j) of AI_n is:

$$(AI_n)_{ij} = \sum_{k=1}^{n} a_{ik} \delta_{kj}.$$

Since $\delta_{kj} = 1$ only when $k = j$, the nonzero summand is a_{ij}, so:

$$(AI_n)_{ij} = a_{ij}.$$

Thus, $AI_n = A$.
Similarly, let I_m be the identity matrix of order $m \times m$. The element at position (i, j) of $I_m A$ is:

$$(I_m A)_{ij} = \sum_{k=1}^{m} \delta_{ik} a_{kj}.$$

Here, $\delta_{ik} = 1$ only when $i = k$, resulting in:

$$(I_m A)_{ij} = a_{ij}.$$

Therefore, $I_m A = A$, proving that the identity matrix acts as the neutral element for matrix multiplication. ∎

> **Corollary 11.1.4** If A is an invertible square matrix of order n, there exists a matrix A^{-1} such that:
>
> $$AA^{-1} = A^{-1}A = I_n$$
>
> The matrix A^{-1} is the **inverse** of A, and the existence of the inverse matrix implies that A has a nonzero determinant.

Demostración. Let A be a square matrix of order n. A is invertible if there exists a matrix A^{-1} of order $n \times n$ such that:

$$AA^{-1} = I_n \quad \text{and} \quad A^{-1}A = I_n,$$

where I_n is the identity matrix of order n.
The existence of such a matrix A^{-1} is both necessary and sufficient for the determinant of A, denoted as $\det(A)$, to be nonzero ($\det(A) \neq 0$). This ensures that the matrix is nonsingular, allowing the construction of A^{-1}.
Therefore, when $\det(A) \neq 0$, there exists a matrix A^{-1} satisfying the above equalities, demonstrating that A has an inverse. ∎

Matrix multiplication, unlike addition, requires the number of columns in the first matrix to equal the number of rows in the second. Additionally, as mentioned earlier, matrix multiplication is not commutative, which is crucial to remember when manipulating algebraic expressions involving matrices.

Exercise 11.3 Compute the product of the following matrices:

$$A = \begin{bmatrix} 1 & 0 & 2 \\ -1 & 3 & 1 \end{bmatrix}, \quad B = \begin{bmatrix} 3 & 1 \\ 2 & 1 \\ 1 & 0 \end{bmatrix}$$

Exercise 11.4 Determine whether the product of matrices A and B, where:

$$A = \begin{bmatrix} 2 & 1 \\ 0 & 3 \end{bmatrix}, \quad B = \begin{bmatrix} 1 & 4 \\ 3 & 2 \end{bmatrix}$$

is equal to the product B times A. Justify your answer.

In this section, we explored the fundamental properties of matrix multiplication, highlighting the importance of associativity and the existence of the identity matrix. We also demonstrated how to compute matrix products and emphasized the non-commutative nature of matrix multiplication.

11.1.3 Transpose of a Matrix

Definition 11.1.3 The **transpose of a matrix** A of order $m \times n$, denoted as A^T, is the matrix obtained by swapping its rows and columns. In other words, the element a_{ij} of the original matrix becomes the element a_{ji} of the transposed matrix. If $A = [a_{ij}]$, then:

$$A^T = [a_{ji}]$$

■ **Example 11.4** Consider the matrix:

$$A = \begin{bmatrix} 1 & 2 & 3 \\ 4 & 5 & 6 \end{bmatrix}$$

The transpose of A, denoted as A^T, is:

$$A^T = \begin{bmatrix} 1 & 4 \\ 2 & 5 \\ 3 & 6 \end{bmatrix}$$

■

Lema 11.1.3 The transpose operation is **involutive**, meaning that applying the transpose operation twice yields the original matrix:

$$(A^T)^T = A$$

11.1 Matrix Operations (Addition, Subtraction, Multiplication)

Demostración. Let $A = [a_{ij}]$ be a matrix. By definition, $A^T = [a_{ji}]$, and applying the transpose again results in $(A^T)^T = [a_{ij}] = A$. Therefore, $(A^T)^T = A$. ∎

Theorem 11.1.5 The transpose of the sum of two matrices is equal to the sum of their transposes. Let A and B be two matrices of the same order, then:

$$(A+B)^T = A^T + B^T$$

This result shows that the transpose operation is linear with respect to the addition of matrices.

Demostración. Let $A = [a_{ij}]$ and $B = [b_{ij}]$. The matrix sum $A + B = [a_{ij} + b_{ij}]$ has a transpose $(A+B)^T = [a_{ji} + b_{ji}]$, which is precisely $A^T + B^T$. Therefore:

$$(A+B)^T = A^T + B^T.$$

∎

Corollary 11.1.6 The transpose of the product of two matrices is equal to the product of their transposes, but in reverse order. Let A and B be matrices such that their product AB is defined, then:

$$(AB)^T = B^T A^T$$

This property is fundamental for understanding the relationship between matrix multiplication and transposition.

Demostración. Let $C = AB$, with $C = [c_{ij}]$, where $c_{ij} = \sum_k a_{ik} b_{kj}$. The transpose of C is $C^T = [c_{ji}] = \left[\sum_k b_{jk} a_{ki}\right]$, which corresponds to the product of B^T and A^T. Thus:

$$(AB)^T = B^T A^T.$$

∎

(R) It is important to note that transposition does not affect the determinant of a square matrix. In other words, for a square matrix A:

$$\det(A) = \det(A^T)$$

Exercise 11.5 Calculate the transpose of the following matrix:

$$B = \begin{bmatrix} 0 & 2 & -1 \\ 3 & 5 & 4 \end{bmatrix}$$

Exercise 11.6 Prove that for the matrices $A = \begin{bmatrix} 1 & 3 \\ 2 & 4 \end{bmatrix}$ and $B = \begin{bmatrix} 5 & 6 \\ 7 & 8 \end{bmatrix}$, the property $(A+B)^T = A^T + B^T$ holds.

In this section, we have explored the definition and fundamental properties of the transpose of a matrix, as well as its role in common operations such as the addition and multiplication of matrices. The properties of transposition simplify many calculations in linear algebra and enhance our understanding of matrix structure.

11.2 Determinant Calculation and Adjoint of a Matrix

11.2.1 Determinants of Matrices of Order 2 and 3

Definition 11.2.1 The **determinant** of a square matrix of order 2 or 3 is a scalar value associated with the matrix that provides information about certain properties, such as whether the matrix is invertible. For a 2×2 matrix A given by:

$$A = \begin{bmatrix} a & b \\ c & d \end{bmatrix}$$

the determinant of A, denoted as $\det(A)$, is calculated as:

$$\det(A) = ad - bc$$

For a 3×3 matrix B given by:

$$B = \begin{bmatrix} a & b & c \\ d & e & f \\ g & h & i \end{bmatrix}$$

the determinant is calculated as:

$$\det(B) = a \begin{vmatrix} e & f \\ h & i \end{vmatrix} - b \begin{vmatrix} d & f \\ g & i \end{vmatrix} + c \begin{vmatrix} d & e \\ g & h \end{vmatrix}$$

■ **Example 11.5** Consider the 2×2 matrix:

$$A = \begin{bmatrix} 2 & 3 \\ 1 & 4 \end{bmatrix}$$

The determinant of A is:

$$\det(A) = (2)(4) - (3)(1) = 8 - 3 = 5$$

For a 3×3 matrix, consider:

$$B = \begin{bmatrix} 1 & 2 & 3 \\ 0 & 4 & 5 \\ 1 & 0 & 6 \end{bmatrix}$$

The determinant of B is:

11.2 Determinant Calculation and Adjoint of a Matrix

$$\det(B) = 1\begin{vmatrix} 4 & 5 \\ 0 & 6 \end{vmatrix} - 2\begin{vmatrix} 0 & 5 \\ 1 & 6 \end{vmatrix} + 3\begin{vmatrix} 0 & 4 \\ 1 & 0 \end{vmatrix}$$

Expanding the minors:

$$\det(B) = (1)(4\cdot 6 - 5\cdot 0) - 2(0\cdot 6 - 5\cdot 1) + 3(0\cdot 0 - 4\cdot 1) = (1)(24) - (2)(-5) + (3)(-4) = 24 + 10 - 12 = 22$$

∎

Lema 11.2.1 The determinant of a triangular matrix (upper or lower triangular) is equal to the product of the elements on the main diagonal. That is, if A is a triangular matrix of order n, then:

$$\det(A) = a_{11}a_{22}\cdots a_{nn}$$

Demostración. Let A be a triangular matrix of order n. Recall that the determinant of a matrix can be computed by expanding along any row or column.
In a triangular matrix, all elements outside the main diagonal are zero. This means that during the determinant expansion, every cofactor corresponding to an off-diagonal element will be multiplied by zero.
As a result, the determinant of A reduces to the product of the diagonal elements, as only terms involving the diagonal contribute to the determinant. Therefore:

$$\det(A) = a_{11}a_{22}\cdots a_{nn}.$$

∎

> **Theorem 11.2.1** A square matrix A is **invertible** if and only if its determinant is nonzero, i.e.:
>
> $$\det(A) \neq 0 \iff A \text{ is invertible.}$$
>
> This result is fundamental, as it allows us to determine whether a matrix is invertible simply by evaluating its determinant.

Demostración. To prove this theorem, consider both implications.
($\Leftarrow$) A is invertible, then $\det(A) \neq 0$: If A is invertible, there exists a matrix A^{-1} such that $AA^{-1} = I$, where I is the identity matrix. Using the property of determinants for a product, we have:

$$\det(AA^{-1}) = \det(I) = 1.$$

Since $\det(AA^{-1}) = \det(A) \cdot \det(A^{-1})$, it follows that $\det(A) \neq 0$ (otherwise, the product would be zero, not 1).
($\Rightarrow$) $\det(A) \neq 0$, then A is invertible: If $\det(A) \neq 0$, the system of linear equations $A\mathbf{x} = \mathbf{b}$ has a unique solution for any vector $\mathbf{b}$, which implies that there exists an inverse matrix A^{-1} satisfying $AA^{-1} = I$. Thus, A is invertible.
Therefore, A is invertible if and only if $\det(A) \neq 0$.

∎

Corollary 11.2.2 If the determinant of a matrix A is zero, then the matrix is **singular** and does not have an inverse. This result follows directly from the previous theorem.

Demostración. From the previous theorem, we know that a square matrix A is invertible if and only if $\det(A) \neq 0$. By contrapositive, if $\det(A) = 0$, then A is not invertible, meaning that A is singular and has no inverse. ∎

(R) The determinant has useful properties such as linearity with respect to rows and invariance under row swaps with a corresponding change in sign. These properties are helpful when calculating determinants of larger matrices.

Exercise 11.7 Compute the determinant of the matrix:

$$C = \begin{bmatrix} 3 & 2 \\ 1 & 5 \end{bmatrix}$$

Exercise 11.8 Determine whether the following matrix is invertible using its determinant:

$$D = \begin{bmatrix} 2 & -1 & 3 \\ 0 & 4 & 5 \\ 1 & -2 & 1 \end{bmatrix}$$

This section has presented the definitions and methods for calculating determinants of 2×2 and 3×3 matrices, as well as fundamental properties linking determinants to the invertibility of matrices.

11.2.2 Determinant Properties

Definition 11.2.2 The **determinant** is a function that associates a scalar to each square matrix, allowing us to evaluate important characteristics of the matrix such as invertibility. The **properties of the determinant** help simplify its computation and understand its behavior under certain transformations of the matrix.

■ **Example 11.6** Consider the matrix A:

$$A = \begin{bmatrix} 1 & 2 \\ 3 & 4 \end{bmatrix}$$

An important property is that if we interchange two rows, the determinant changes its sign. Let's interchange rows 1 and 2 of A:

$$A' = \begin{bmatrix} 3 & 4 \\ 1 & 2 \end{bmatrix}$$

Then:

$$\det(A) = -\det(A') = (1)(4) - (2)(3) = 4 - 6 = -2$$

Lema 11.2.2 If a matrix A has a row or column of zeros, then $\det(A) = 0$. This property follows from the fact that expanding the determinant along a row or column of zeros results in all terms being null.

Demostración. Suppose A is a matrix with a row (or column) of zeros. When calculating the determinant by expanding along that row (or column), every term of the expansion includes an element from that row (or column) which is zero. Thus, all terms of the expansion are zero, and therefore:

$$\det(A) = 0.$$

■

Theorem 11.2.3 For a square matrix A, the determinant is **linear** with respect to each of its rows or columns. That is, if a row (or column) of A can be expressed as a linear combination of two vectors, the determinant of A is also the linear combination of the determinants of the respective matrices.

Demostración. Suppose a row of A can be expressed as a linear combination of two vectors $\mathbf{v}$ and $\mathbf{w}$, i.e., there exists a scalar α such that:

$$\text{row}_i(A) = \alpha \mathbf{v} + \mathbf{w}.$$

We can express A as a linear combination of two matrices A_1 and A_2, where A_1 has the i-th row equal to $\alpha \mathbf{v}$ and A_2 has the i-th row equal to $\mathbf{w}$, with all other rows identical to those of A. By the linearity property of the determinant with respect to a row, we have:

$$\det(A) = \det(A_1 + A_2) = \alpha \det(A_1) + \det(A_2).$$

Thus, the determinant of A is a linear combination of the determinants of A_1 and A_2, which proves the linearity of the determinant with respect to a row (or column). ■

Corollary 11.2.4 If a matrix A has two proportional rows or columns, then its determinant is zero. This is because having proportional rows or columns implies that the matrix is linearly dependent and therefore not invertible.

Demostración. Suppose A has two proportional rows, i.e., there exists a scalar α such that one row is α times another row. Since the rows are linearly dependent, the matrix A does not have full rank and is not invertible. By the invertibility theorem, this implies that $\det(A) = 0$. ■

> **R** The determinant also has the **multiplicative property**: the determinant of the product of two matrices is equal to the product of their determinants. That is, if A and B are square matrices of order n, then:
>
> $$\det(AB) = \det(A) \cdot \det(B)$$
>
> This property is useful in linear algebra and practical applications.

Exercise 11.9 Prove that if A is a triangular matrix of order 3×3, the determinant is equal to the product of the elements on the main diagonal.

Exercise 11.10 Verify if the matrix B given by:

$$B = \begin{bmatrix} 2 & 4 & 6 \\ 1 & 2 & 3 \\ 3 & 6 & 9 \end{bmatrix}$$

has a determinant equal to zero, and explain why using determinant properties.

These results and properties help us better understand how determinants are calculated and how they behave under various operations, enabling us to assess invertibility and other important characteristics of matrices.

11.2.3 Adjoint of a Matrix

Definition 11.2.3 The **adjoint** of a square matrix A of order $n \times n$ is defined as the transpose of the **cofactor matrix** of A. Each element of the cofactor matrix is obtained by calculating the determinant of the corresponding minor with the appropriate sign.

■ **Example 11.7** Consider the matrix:

$$A = \begin{bmatrix} 2 & 1 \\ 3 & 4 \end{bmatrix}$$

First, calculate the cofactors:

$$C_{11} = \det[4] = 4, \quad C_{12} = -\det[3] = -3$$
$$C_{21} = -\det[1] = -1, \quad C_{22} = \det[2] = 2$$

Thus, the cofactor matrix is:

$$C = \begin{bmatrix} 4 & -3 \\ -1 & 2 \end{bmatrix}$$

The adjoint of A, denoted as $\mathrm{adj}(A)$, is the transpose of C:

$$\mathrm{adj}(A) = \begin{bmatrix} 4 & -1 \\ -3 & 2 \end{bmatrix}$$

Lema 11.2.3 For a square matrix A of order $n \times n$, the product of the matrix A and its adjoint $\mathrm{adj}(A)$ is a multiple of the determinant of A and the identity matrix of order n:

$$A \cdot \mathrm{adj}(A) = \det(A) \cdot I_n$$

This means that if A is invertible, its adjoint can be used to calculate the inverse matrix.

Demostración. The adjoint $\mathrm{adj}(A)$ of A is defined such that each element of $A \cdot \mathrm{adj}(A)$ is the corresponding cofactor multiplied by the element of A, resulting in the determinant expanded along each row of A. This produces a diagonal matrix with $\det(A)$ on the diagonal entries and zeros elsewhere. Therefore:

$$A \cdot \mathrm{adj}(A) = \det(A) \cdot I_n.$$

11.2 Determinant Calculation and Adjoint of a Matrix

Theorem 11.2.5 If the determinant of a square matrix A is non-zero, then the matrix is **invertible**, and its inverse can be calculated using the adjoint:

$$A^{-1} = \frac{1}{\det(A)} \operatorname{adj}(A)$$

This relationship is fundamental for finding the inverse of a matrix when its determinant and adjoint are known.

Demostración. If $\det(A) \neq 0$, the product of A and its adjoint satisfies:

$$A \cdot \operatorname{adj}(A) = \det(A) \cdot I_n.$$

Dividing both sides by $\det(A)$, we obtain:

$$A \cdot \frac{1}{\det(A)} \operatorname{adj}(A) = I_n.$$

This shows that $\frac{1}{\det(A)} \operatorname{adj}(A)$ is the inverse of A, so:

$$A^{-1} = \frac{1}{\det(A)} \operatorname{adj}(A).$$

∎

Corollary 11.2.6 If a matrix A is singular (i.e., its determinant is zero), then its inverse does not exist. This follows from the fact that division by zero is undefined in the formula above.

Demostración. If A is singular, then $\det(A) = 0$. In this case, the formula for the inverse,

$$A^{-1} = \frac{1}{\det(A)} \operatorname{adj}(A),$$

is not applicable as it involves division by zero. Therefore, the inverse of A does not exist when $\det(A) = 0$.

∎

> (R) The calculation of the adjoint of a matrix can be simplified using determinant properties such as linearity and Sarrus's rule (for 3×3 matrices). These properties help avoid common errors when computing cofactors.

Exercise 11.11 Calculate the adjoint of the matrix:

$$B = \begin{bmatrix} 1 & 0 & 2 \\ -1 & 3 & 1 \\ 2 & 4 & 0 \end{bmatrix}$$

and verify if the relation $B \cdot \operatorname{adj}(B) = \det(B) \cdot I_3$ holds.

Exercise 11.12 Find the inverse of the matrix:

$$C = \begin{bmatrix} 3 & 2 \\ 1 & 4 \end{bmatrix}$$

using the formula involving the adjoint, and verify that the inverse is correct by computing the product $C \cdot C^{-1}$. ∎

These concepts and results regarding the adjoint of a matrix are essential for understanding the process of matrix inversion, as well as solving systems of linear equations using Cramer's rule and other algebraic methods.

11.3 Inverse of a Matrix and Its Application in Systems of Equations

11.3.1 Definition of the Inverse Matrix

Definition 11.3.1 Let A be a square matrix of order $n \times n$. The matrix A is said to be **invertible** if there exists a matrix B of the same order such that:

$$A \cdot B = B \cdot A = I_n$$

where I_n is the identity matrix of order n. This matrix B is called the **inverse matrix** of A, and it is denoted as A^{-1}.

■ **Example 11.8** Consider the matrix:

$$A = \begin{bmatrix} 2 & 1 \\ 3 & 4 \end{bmatrix}$$

To find the inverse matrix A^{-1}, first calculate the determinant of A:

$$\det(A) = 2 \cdot 4 - 1 \cdot 3 = 8 - 3 = 5$$

Since the determinant is non-zero, the matrix is invertible. The inverse is calculated as:

$$A^{-1} = \frac{1}{\det(A)} \begin{bmatrix} 4 & -1 \\ -3 & 2 \end{bmatrix} = \frac{1}{5} \begin{bmatrix} 4 & -1 \\ -3 & 2 \end{bmatrix}$$

Thus:

$$A^{-1} = \begin{bmatrix} \frac{4}{5} & -\frac{1}{5} \\ -\frac{3}{5} & \frac{2}{5} \end{bmatrix}$$

∎

Lema 11.3.1 If A and B are square matrices of order $n \times n$ and A is invertible, then the inverse of the product $A \cdot B$ is equal to $B^{-1} \cdot A^{-1}$, provided B is also invertible. In symbols:

$$(A \cdot B)^{-1} = B^{-1} \cdot A^{-1}$$

11.3 Inverse of a Matrix and Its Application in Systems of Equations

Demostración. To prove that $(A \cdot B)^{-1} = B^{-1} \cdot A^{-1}$, we verify that the product $(A \cdot B) \cdot (B^{-1} \cdot A^{-1})$ results in the identity matrix. Multiplying, we obtain:

$$(A \cdot B) \cdot (B^{-1} \cdot A^{-1}) = A \cdot (B \cdot B^{-1}) \cdot A^{-1} = A \cdot I \cdot A^{-1} = A \cdot A^{-1} = I.$$

Similarly, multiplying in the reverse order:

$$(B^{-1} \cdot A^{-1}) \cdot (A \cdot B) = B^{-1} \cdot (A^{-1} \cdot A) \cdot B = B^{-1} \cdot I \cdot B = B^{-1} \cdot B = I.$$

Thus, $(A \cdot B)^{-1} = B^{-1} \cdot A^{-1}$. ∎

> **Theorem 11.3.1** A square matrix A of order $n \times n$ is invertible if and only if its determinant is non-zero. In other words:
>
> $$A^{-1} \text{ exists} \iff \det(A) \neq 0$$
>
> This condition is both necessary and sufficient, meaning that only matrices with a non-zero determinant have an inverse.

Demostración. To prove this theorem, we show both directions of the implication.
1. **(If A is invertible, then $\det(A) \neq 0$):** If A is invertible, there exists a matrix A^{-1} such that $A \cdot A^{-1} = I$, where I is the identity matrix. Using the determinant property of products, we have:

$$\det(A \cdot A^{-1}) = \det(I) = 1.$$

Since $\det(A \cdot A^{-1}) = \det(A) \cdot \det(A^{-1})$, it follows that $\det(A) \neq 0$ (otherwise, the product would be zero, which contradicts it being 1).
2. **(If $\det(A) \neq 0$, then A is invertible):** If $\det(A) \neq 0$, the system of linear equations $A\mathbf{x} = \mathbf{b}$ has a unique solution for any vector $\mathbf{b}$, implying the existence of an inverse matrix A^{-1} such that $A \cdot A^{-1} = I$. Hence, A is invertible.
In conclusion, A is invertible if and only if $\det(A) \neq 0$. ∎

> **Corollary 11.3.2** If A and B are square matrices of order $n \times n$ and $A \cdot B = I_n$, then A and B are inverses of each other, meaning $A = B^{-1}$ and $B = A^{-1}$.

Demostración. Given that $A \cdot B = I_n$, B acts as a right inverse of A. Similarly, if we multiply both sides by A on the right, we get $B \cdot A = I_n$, showing that A also acts as a right inverse of B.
By definition, a square matrix of order $n \times n$ that has a right inverse and a left inverse is invertible, and the inverse is unique. Hence, $A = B^{-1}$ and $B = A^{-1}$. ∎

> (R) The calculation of the inverse matrix can be simplified using the reduced row echelon form or the adjoint method, depending on the size of the matrix and the context in which it is applied. Both methods require the determinant to be non-zero to ensure the existence of the inverse.

> **Exercise 11.13** Find the inverse of the matrix:
>
> $$B = \begin{bmatrix} 1 & 2 \\ 3 & 5 \end{bmatrix}$$

and verify that the product $B \cdot B^{-1}$ equals the identity matrix of order 2×2.

Exercise 11.14 Determine whether the matrix:

$$C = \begin{bmatrix} 2 & 4 \\ 1 & 2 \end{bmatrix}$$

is invertible, and if so, calculate its inverse.

These definitions and results on the inverse of a matrix are fundamental for solving systems of linear equations using the matrix method, as well as for understanding advanced concepts in linear algebra and their applications.

11.3.2 Method for Calculating the Inverse (Gauss-Jordan)

Definition 11.3.2 The **Gauss-Jordan method** for finding the inverse of a matrix A of order $n \times n$ involves applying row elementary operations to the augmented matrix $[A|I_n]$, where I_n is the identity matrix of order n. The goal is to transform A into the identity matrix, and the resulting matrix on the right will be the inverse A^{-1}, provided A is invertible.

■ **Example 11.9** Consider the matrix:

$$A = \begin{bmatrix} 1 & 2 \\ 3 & 4 \end{bmatrix}$$

To find the inverse of A using the Gauss-Jordan method, start with the augmented matrix $[A|I]$:

$$\left[\begin{array}{cc|cc} 1 & 2 & 1 & 0 \\ 3 & 4 & 0 & 1 \end{array} \right]$$

Apply row operations to transform A into the identity matrix:
1. Subtract 3 times the first row from the second row to eliminate the element 3 in the first column of the second row:

$$\left[\begin{array}{cc|cc} 1 & 2 & 1 & 0 \\ 0 & -2 & -3 & 1 \end{array} \right]$$

2. Multiply the second row by $-\frac{1}{2}$:

$$\left[\begin{array}{cc|cc} 1 & 2 & 1 & 0 \\ 0 & 1 & \frac{3}{2} & -\frac{1}{2} \end{array} \right]$$

3. Subtract 2 times the second row from the first row to make the element 2 equal to 0:

$$\left[\begin{array}{cc|cc} 1 & 0 & -2 & 1 \\ 0 & 1 & \frac{3}{2} & -\frac{1}{2} \end{array} \right]$$

Thus, the inverse matrix is:

$$A^{-1} = \begin{bmatrix} -2 & 1 \\ \frac{3}{2} & -\frac{1}{2} \end{bmatrix}$$

11.3 Inverse of a Matrix and Its Application in Systems of Equations

Lema 11.3.2 If a matrix A of order $n \times n$ has a non-zero determinant, then the Gauss-Jordan method guarantees the existence and uniqueness of its inverse matrix.

Demostración. If $\det(A) \neq 0$, then A is invertible. The Gauss-Jordan method transforms A into the identity matrix I using row elementary operations, which are invertible and preserve the existence of the inverse. When the Gauss-Jordan method is applied to the augmented matrix $(A \mid I)$, these operations transform A into I and I into A^{-1}.

Since the determinant is non-zero, all rows of A are linearly independent, ensuring that the method will reach the identity matrix and produce a unique matrix A^{-1} in the process. Thus, the existence and uniqueness of the inverse are guaranteed. ∎

Theorem 11.3.3 The Gauss-Jordan method is equivalent to solving the system of equations $A\mathbf{x} = \mathbf{b}$ for all columns of the vector $\mathbf{b}$ in the identity matrix I_n. By applying this process to the augmented matrix $[A|I_n]$, the inverse of A is obtained, if it exists.

Demostración. To find the inverse of A using Gauss-Jordan, consider the augmented matrix $[A|I_n]$, where I_n is the identity matrix of order n. Applying row elementary operations to bring A to the form of I_n, these same operations simultaneously transform I_n into a new matrix denoted as B.

If A is invertible, the row operations will convert A into I_n, and the matrix B obtained in the process will satisfy $AB = I_n$. This shows that $B = A^{-1}$. Thus, the Gauss-Jordan method applied to $[A|I_n]$ produces A^{-1} when $\det(A) \neq 0$. ∎

Corollary 11.3.4 If a square matrix A can be reduced to the identity matrix using the Gauss-Jordan method, then A is invertible, and the matrix accompanying the identity in the process is precisely A^{-1}.

Demostración. Applying the Gauss-Jordan method to the augmented matrix $[A|I_n]$, if A is reduced to the identity I_n through row elementary operations, these operations simultaneously transform I_n into a matrix B. Since A has been transformed into I_n, the matrix B must satisfy $AB = I_n$, hence $B = A^{-1}$.

Therefore, if A is reduced to I_n, A is invertible, and B is its inverse A^{-1}. ∎

> **R** The Gauss-Jordan method is particularly efficient for small matrices. For larger matrices, other methods, such as LU decomposition or numerical methods, are often preferred due to their computational stability.

Exercise 11.15 Find the inverse of the matrix:

$$B = \begin{bmatrix} 2 & 3 \\ 1 & 2 \end{bmatrix}$$

using the Gauss-Jordan method.

Exercise 11.16 Determine whether the matrix:

$$C = \begin{bmatrix} 1 & 0 & 2 \\ 0 & 1 & 3 \\ 4 & 5 & 6 \end{bmatrix}$$

is invertible by applying the Gauss-Jordan method. If it is, calculate its inverse.

These concepts and examples on the Gauss-Jordan method provide a clear understanding of how this procedure can be used to find the inverse of a matrix, as well as its relevance and practical limitations in problem-solving.

11.3.3 Applications of the Inverse Matrix in Solving Systems

Definition 11.3.3 The **inverse matrix** of a square matrix A, denoted as A^{-1}, is used to solve linear systems of equations of the form $Ax = b$. If A is invertible, the solution can be expressed as $\mathbf{x} = A^{-1}\mathbf{b}$.

■ **Example 11.10** Consider the system of equations:

$$\begin{cases} 2x + 3y = 5 \\ x + 4y = 6 \end{cases}$$

We can write this in matrix form as:

$$A\mathbf{x} = \mathbf{b}, \quad \text{where} \quad A = \begin{bmatrix} 2 & 3 \\ 1 & 4 \end{bmatrix}, \quad \mathbf{x} = \begin{bmatrix} x \\ y \end{bmatrix}, \quad \mathbf{b} = \begin{bmatrix} 5 \\ 6 \end{bmatrix}$$

First, calculate the inverse of A:

$$A^{-1} = \frac{1}{(2)(4) - (3)(1)} \begin{bmatrix} 4 & -3 \\ -1 & 2 \end{bmatrix} = \frac{1}{5} \begin{bmatrix} 4 & -3 \\ -1 & 2 \end{bmatrix}$$

Then, the solution is:

$$\mathbf{x} = A^{-1}\mathbf{b} = \frac{1}{5} \begin{bmatrix} 4 & -3 \\ -1 & 2 \end{bmatrix} \begin{bmatrix} 5 \\ 6 \end{bmatrix} = \frac{1}{5} \begin{bmatrix} 20 - 18 \\ -5 + 12 \end{bmatrix} = \begin{bmatrix} \frac{2}{5} \\ \frac{7}{5} \end{bmatrix}$$

Thus, the solution to the system is $x = \frac{2}{5}$ and $y = \frac{7}{5}$. ■

Lema 11.3.3 If a matrix A is invertible, then the linear system $Ax = b$ has a unique solution given by $\mathbf{x} = A^{-1}\mathbf{b}$. This result allows us to determine the uniqueness of the solution of a system using the inverse matrix.

Demostración. Since A is invertible, there exists a matrix A^{-1} such that $AA^{-1} = I$, where I is the identity matrix. To solve $Ax = b$, multiply both sides of the equation by A^{-1}:

$$A^{-1}A\mathbf{x} = A^{-1}\mathbf{b}.$$

This simplifies to:

$$I\mathbf{x} = A^{-1}\mathbf{b},$$

so $\mathbf{x} = A^{-1}\mathbf{b}$ is the unique solution of the system.

Uniqueness is guaranteed since A is invertible, meaning no other solution exists for $Ax = b$. ∎

11.3 Inverse of a Matrix and Its Application in Systems of Equations

Theorem 11.3.5 If A is an invertible matrix of order $n \times n$, then the homogeneous system $A\mathbf{x} = \mathbf{0}$ has the unique solution $\mathbf{x} = \mathbf{0}$. This is because multiplying both sides by A^{-1} yields $A^{-1}A\mathbf{x} = A^{-1}\mathbf{0} \Rightarrow \mathbf{x} = \mathbf{0}$.

Demostración. Given the homogeneous system $A\mathbf{x} = \mathbf{0}$ and the fact that A is invertible, multiply both sides of the equation by A^{-1}:

$$A^{-1}A\mathbf{x} = A^{-1}\mathbf{0}.$$

This simplifies to:

$$I\mathbf{x} = \mathbf{0},$$

which implies that $\mathbf{x} = \mathbf{0}$.
Thus, $\mathbf{x} = \mathbf{0}$ is the only solution to the system $A\mathbf{x} = \mathbf{0}$, given that A is invertible. ∎

Corollary 11.3.6 If the determinant of a square matrix A is nonzero, then A is invertible, and therefore, the system $A\mathbf{x} = \mathbf{b}$ has a unique solution. This corollary reinforces the connection between the determinant and the existence of a unique solution in linear systems.

Demostración. If $\det(A) \neq 0$, then A is invertible. This implies the existence of an inverse matrix A^{-1} such that $AA^{-1} = I$.
To solve the system $A\mathbf{x} = \mathbf{b}$, multiply both sides by A^{-1}:

$$A^{-1}A\mathbf{x} = A^{-1}\mathbf{b}.$$

This simplifies to:

$$I\mathbf{x} = A^{-1}\mathbf{b},$$

giving the unique solution:

$$\mathbf{x} = A^{-1}\mathbf{b}.$$

Thus, the system has a unique solution when $\det(A) \neq 0$. ∎

> **R** The application of the inverse matrix to solve linear systems is practical for small matrices. However, for large matrices, using the inverse matrix can be computationally expensive and prone to numerical errors. In such cases, other methods, such as *LU* factorization or Gaussian elimination, are preferable.

Exercise 11.17 Solve the linear system using the inverse matrix:

$$\begin{cases} 3x + 4y = 7 \\ 5x + 2y = 8 \end{cases}$$

Exercise 11.18 Given the matrix $A = \begin{bmatrix} 1 & 2 & 3 \\ 0 & 1 & 4 \\ 5 & 6 & 0 \end{bmatrix}$, verify if the matrix is invertible by calculating its determinant. If it is invertible, find its inverse and use it to solve the system $A\mathbf{x} = \mathbf{b}$, where

$$\mathbf{b} = \begin{bmatrix} 1 \\ 2 \\ 3 \end{bmatrix}.$$

This development provides a comprehensive view of the use of the inverse matrix in solving linear systems, highlighting both the theoretical aspects and the practical application of this fundamental concept in linear algebra.

11.4 Ejercicios Resueltos

Exercise 11.19 Calcula el determinante de la matriz $A = \begin{bmatrix} 2 & -1 \\ 3 & 4 \end{bmatrix}$.

Demostración. El determinante de A se calcula como:

$$\det(A) = (2)(4) - (-1)(3) = 8 + 3 = 11.$$

∎

Exercise 11.20 Suma las matrices $A = \begin{bmatrix} 1 & 2 \\ 3 & 4 \end{bmatrix}$ y $B = \begin{bmatrix} 4 & 3 \\ 2 & 1 \end{bmatrix}$.

Demostración. La suma de matrices $A + B$ es:

$$A + B = \begin{bmatrix} 1+4 & 2+3 \\ 3+2 & 4+1 \end{bmatrix} = \begin{bmatrix} 5 & 5 \\ 5 & 5 \end{bmatrix}.$$

∎

Exercise 11.21 Determina si la matriz $B = \begin{bmatrix} 1 & 0 & 3 \\ 2 & 1 & 4 \\ 3 & -1 & 5 \end{bmatrix}$ es invertible.

Demostración. Para determinar si una matriz es invertible, necesitamos calcular su determinante. Si el determinante es diferente de cero, la matriz es invertible. Calculando el determinante de B, obtenemos:

$$\det(B) = 1(1 \cdot 5 - 4 \cdot (-1)) - 0 + 3(2 \cdot (-1) - 3 \cdot 1) = 1(5+4) + 3(-2-3) = 9 - 15 = -6 \neq 0.$$

Por lo tanto, la matriz B es invertible. ∎

Exercise 11.22 Encuentra el producto de las matrices $A = \begin{bmatrix} 1 & 2 \\ 3 & 4 \end{bmatrix}$ y $B = \begin{bmatrix} 2 & 0 \\ 1 & 3 \end{bmatrix}$.

Demostración. El producto AB es:

$$AB = \begin{bmatrix} 1 \cdot 2 + 2 \cdot 1 & 1 \cdot 0 + 2 \cdot 3 \\ 3 \cdot 2 + 4 \cdot 1 & 3 \cdot 0 + 4 \cdot 3 \end{bmatrix} = \begin{bmatrix} 4 & 6 \\ 10 & 12 \end{bmatrix}.$$

∎

11.5 Solved Exercises

Exercise 11.23 Resuelve el sistema de ecuaciones representado por la matriz aumentada usando el método de eliminación por filas:

$$\left[\begin{array}{ccc|c} 1 & 2 & -1 & 3 \\ 0 & 1 & 3 & 4 \\ 2 & -3 & 4 & -5 \end{array}\right]$$

Demostración. Realizando operaciones de fila: 1. $F_3 \to F_3 - 2F_1$:

$$\left[\begin{array}{ccc|c} 1 & 2 & -1 & 3 \\ 0 & 1 & 3 & 4 \\ 0 & -7 & 6 & -11 \end{array}\right]$$

2. $F_3 \to F_3 + 7F_2$:

$$\left[\begin{array}{ccc|c} 1 & 2 & -1 & 3 \\ 0 & 1 & 3 & 4 \\ 0 & 0 & 27 & 17 \end{array}\right]$$

El sistema resuelto es:

$$x = -1, \quad y = 4, \quad z = \frac{17}{27}$$

∎

11.5 Solved Exercises

Exercise 11.24 Calculate the determinant of the matrix $A = \begin{bmatrix} 2 & -1 \\ 3 & 4 \end{bmatrix}$.

Demostración. The determinant of A is calculated as:

$$\det(A) = (2)(4) - (-1)(3) = 8 + 3 = 11.$$

∎

Exercise 11.25 Add the matrices $A = \begin{bmatrix} 1 & 2 \\ 3 & 4 \end{bmatrix}$ and $B = \begin{bmatrix} 4 & 3 \\ 2 & 1 \end{bmatrix}$.

Demostración. The sum of the matrices $A + B$ is:

$$A + B = \begin{bmatrix} 1+4 & 2+3 \\ 3+2 & 4+1 \end{bmatrix} = \begin{bmatrix} 5 & 5 \\ 5 & 5 \end{bmatrix}.$$

∎

Exercise 11.26 Determine whether the matrix $B = \begin{bmatrix} 1 & 0 & 3 \\ 2 & 1 & 4 \\ 3 & -1 & 5 \end{bmatrix}$ is invertible.

Demostración. To determine if a matrix is invertible, we calculate its determinant. If the determinant is nonzero, the matrix is invertible. Calculating the determinant of B, we have:

$$\det(B) = 1(1 \cdot 5 - 4 \cdot (-1)) - 0 + 3(2 \cdot (-1) - 3 \cdot 1) = 1(5+4) + 3(-2-3) = 9 - 15 = -6 \neq 0.$$

Thus, the matrix B is invertible. ∎

Exercise 11.27 Find the product of the matrices $A = \begin{bmatrix} 1 & 2 \\ 3 & 4 \end{bmatrix}$ and $B = \begin{bmatrix} 2 & 0 \\ 1 & 3 \end{bmatrix}$.

Demostración. The product AB is:

$$AB = \begin{bmatrix} 1 \cdot 2 + 2 \cdot 1 & 1 \cdot 0 + 2 \cdot 3 \\ 3 \cdot 2 + 4 \cdot 1 & 3 \cdot 0 + 4 \cdot 3 \end{bmatrix} = \begin{bmatrix} 4 & 6 \\ 10 & 12 \end{bmatrix}.$$

∎

Exercise 11.28 Solve the system of equations represented by the augmented matrix using row elimination:

$$\begin{bmatrix} 1 & 2 & -1 & | & 3 \\ 0 & 1 & 3 & | & 4 \\ 2 & -3 & 4 & | & -5 \end{bmatrix}$$

Demostración. Performing row operations: 1. $F_3 \to F_3 - 2F_1$:

$$\begin{bmatrix} 1 & 2 & -1 & | & 3 \\ 0 & 1 & 3 & | & 4 \\ 0 & -7 & 6 & | & -11 \end{bmatrix}$$

2. $F_3 \to F_3 + 7F_2$:

$$\begin{bmatrix} 1 & 2 & -1 & | & 3 \\ 0 & 1 & 3 & | & 4 \\ 0 & 0 & 27 & | & 17 \end{bmatrix}$$

The solved system is:

$$x = -1, \quad y = 4, \quad z = \frac{17}{27}$$

∎

11.6 Proposed Exercises

11.6.1 Matrix Operations (Addition, Subtraction, Multiplication)

Exercise 11.29 Add the matrices $A = \begin{bmatrix} 2 & 4 \\ 1 & 3 \end{bmatrix}$ and $B = \begin{bmatrix} 1 & 2 \\ 3 & 4 \end{bmatrix}$.

11.6 Proposed Exercises

Exercise 11.30 Calculate the subtraction $A - B$ where $A = \begin{bmatrix} 5 & 6 \\ 7 & 8 \end{bmatrix}$ and $B = \begin{bmatrix} 2 & 3 \\ 4 & 1 \end{bmatrix}$.

Exercise 11.31 Find the product of $A = \begin{bmatrix} 1 & 0 \\ 2 & 1 \end{bmatrix}$ by $B = \begin{bmatrix} 3 & 2 \\ 1 & 4 \end{bmatrix}$.

Exercise 11.32 If $A = \begin{bmatrix} 2 & -1 \\ 0 & 3 \end{bmatrix}$ and $B = \begin{bmatrix} 1 & 4 \\ 2 & -3 \end{bmatrix}$, find $A \cdot B$.

Exercise 11.33 Multiply the matrix $C = \begin{bmatrix} 3 & 2 \\ 1 & 5 \end{bmatrix}$ by the scalar $k = 4$.

11.6.2 Determinant and Adjoint Calculation of a Matrix

Exercise 11.34 Calculate the determinant of the matrix $A = \begin{bmatrix} 3 & 5 \\ 2 & 1 \end{bmatrix}$.

Exercise 11.35 Find the determinant of the matrix $B = \begin{bmatrix} 1 & 2 & 3 \\ 0 & 1 & 4 \\ 5 & 6 & 0 \end{bmatrix}$.

Exercise 11.36 Determine the adjoint of the matrix $C = \begin{bmatrix} 2 & 1 \\ 4 & 3 \end{bmatrix}$.

Exercise 11.37 Calculate the determinant of the matrix $D = \begin{bmatrix} 7 & 8 & 9 \\ 4 & 5 & 6 \\ 1 & 2 & 3 \end{bmatrix}$.

Exercise 11.38 Find the adjoint of the matrix $E = \begin{bmatrix} 1 & 0 \\ 3 & 4 \end{bmatrix}$.

11.6.3 Inverse of a Matrix and Its Application in Systems of Equations

Exercise 11.39 Calculate the inverse of the matrix $A = \begin{bmatrix} 2 & 1 \\ 3 & 4 \end{bmatrix}$.

Exercise 11.40 Determine whether the matrix $B = \begin{bmatrix} 1 & 2 \\ 2 & 4 \end{bmatrix}$ has an inverse and justify your answer.

Exercise 11.41 Solve the system of equations $A\mathbf{x} = \mathbf{b}$ using the inverse matrix, where $A = \begin{bmatrix} 3 & 1 \\ 2 & 4 \end{bmatrix}$ and $\mathbf{b} = \begin{bmatrix} 5 \\ 6 \end{bmatrix}$.

Exercise 11.42 Calculate the inverse of the matrix $C = \begin{bmatrix} 4 & 7 \\ 2 & 6 \end{bmatrix}$, if it exists.

Exercise 11.43 Determine the solution of the system of equations using the inverse matrix $D = \begin{bmatrix} 1 & 2 \\ 3 & 5 \end{bmatrix}$ for $\mathbf{b} = \begin{bmatrix} 4 \\ 7 \end{bmatrix}$.

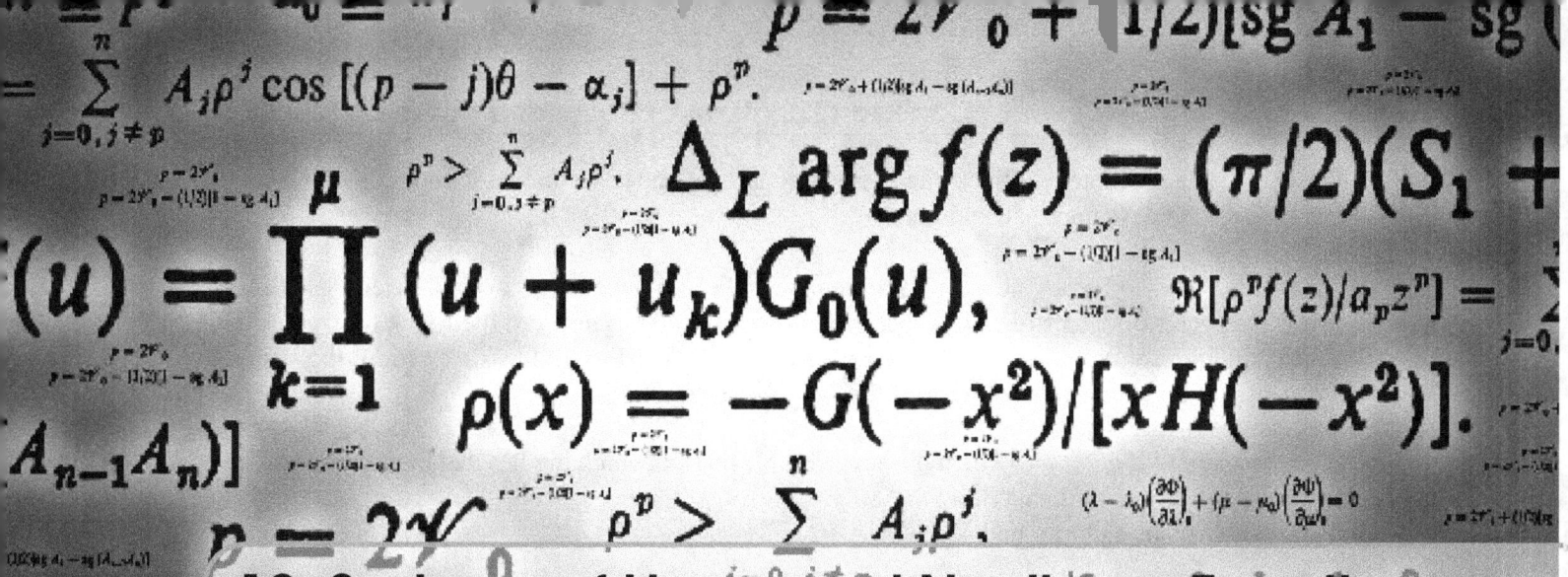

12. Systems of Linear and Nonlinear Equations

12.1 Methods for Solving Linear Systems

12.1.1 Substitution Method and Examples

Definition 12.1.1 The **substitution method** is a technique for solving systems of linear equations that involves isolating one variable in one equation and then substituting its value into the other equation(s) in the system. This reduces the number of unknowns until a solution is obtained. This method is particularly effective for systems with two equations and two unknowns.

■ **Example 12.1** Consider the system of equations:

$$\begin{cases} x + 2y = 10 \\ 3x - y = 5 \end{cases}$$

Applying the substitution method:
1. Solve for x from the first equation:

$$x = 10 - 2y$$

2. Substitute x into the second equation:

$$3(10 - 2y) - y = 5$$

3. Solve for y:

$$30 - 6y - y = 5 \Rightarrow -7y = -25 \Rightarrow y = \frac{25}{7}$$

4. Substitute the value of y into the expression for x:

$$x = 10 - 2\left(\frac{25}{7}\right) = \frac{70 - 50}{7} = \frac{20}{7}$$

Thus, the solution to the system is $x = \frac{20}{7}$, $y = \frac{25}{7}$. ∎

Capítulo 12. Systems of Linear and Nonlinear Equations

Lema 12.1.1 If a system of linear equations is consistent and has a unique solution, the substitution method will always find that solution.

> **Theorem 12.1.1** For a system of linear equations $Ax = b$, where A is the coefficient matrix and b is the vector of independent terms, if the determinant of A is nonzero ($\det(A) \neq 0$), the system has a unique solution. The substitution method can be used to find this solution if the system consists of equations in simple forms for isolating a variable.

Demostración. If $\det(A) \neq 0$, then A is an invertible matrix, which implies that there exists a unique inverse matrix A^{-1} such that $AA^{-1} = I$.
To solve $Ax = b$, multiply both sides by A^{-1}:

$$A^{-1}Ax = A^{-1}b.$$

This simplifies to:

$$Ix = A^{-1}b,$$

resulting in the unique solution:

$$x = A^{-1}b.$$

When the equations in the system have a straightforward structure, the substitution method can be applied to isolate variables in a logical sequence, leading to the same unique solution obtained via the inverse matrix. Therefore, if $\det(A) \neq 0$, the system has a unique solution. ∎

> **Corollary 12.1.2** If the determinant of the coefficient matrix of a system of linear equations is zero, the system may be inconsistent or have infinitely many solutions. In this case, the substitution method may lead to a contradiction or redundant equations.

Demostración. If the determinant of the coefficient matrix A is zero ($\det(A) = 0$), then A is not invertible, which implies that the system $Ax = b$ does not have a unique solution.
In this case, two scenarios may occur: 1. **The system is inconsistent:** No solution exists because the equations lead to a contradiction when attempting to solve them.
2. **The system has infinitely many solutions:** The equations are linearly dependent, meaning at least one equation is a linear combination of the others, resulting in redundant equations.
Therefore, when $\det(A) = 0$, the system lacks a unique solution and may either be inconsistent or have infinitely many solutions. ∎

> (R) The substitution method is recommended when it is straightforward to isolate one variable and substitute it into the other equations. For larger or more complex systems, methods such as Gaussian elimination or matrix methods are generally more efficient.

> **Exercise 12.1** Use the substitution method to solve the following system of equations:
> $$\begin{cases} 2x + 3y = 14 \\ x - 4y = -2 \end{cases}$$

12.1 Methods for Solving Linear Systems

> **Exercise 12.2** Solve the system of equations using the substitution method:
>
> $$\begin{cases} 5x - y = 11 \\ 3x + 2y = 1 \end{cases}$$
>
> Verify your answer by substituting the values of x and y into both equations.

The substitution method is a fundamental approach for solving linear systems, and the examples and exercises provided here will help you better understand how to apply this method in various situations.

12.1.2 Step-by-Step Equalization Method

> **Definition 12.1.2** The **equalization method** is a procedure for solving systems of linear equations where the same variable is isolated in both equations and the resulting expressions are then set equal to each other. This reduces the system to a single equation with one unknown, which can be solved. After finding the value of the first unknown, it is substituted into one of the original equations to determine the value of the other unknown.

■ **Example 12.2** Consider the system of equations:

$$\begin{cases} x + 3y = 7 \\ 2x - y = 1 \end{cases}$$

Applying the equalization method:
1. Solve for x in the first equation:

$$x = 7 - 3y$$

2. Solve for x in the second equation:

$$x = \frac{y+1}{2}$$

3. Set the two expressions equal to each other:

$$7 - 3y = \frac{y+1}{2}$$

4. Solve for y:

$$14 - 6y = y + 1 \Rightarrow 14 - 1 = 7y \Rightarrow 13 = 7y \Rightarrow y = \frac{13}{7}$$

5. Substitute the value of y into one of the equations to find x:

$$x = 7 - 3\left(\frac{13}{7}\right) = \frac{49 - 39}{7} = \frac{10}{7}$$

Thus, the solution to the system is $x = \frac{10}{7}, y = \frac{13}{7}$. ■

Lema 12.1.2 The equalization method can be applied to any system of two linear equations as long as both equations can isolate the same variable without introducing undefined expressions.

Capítulo 12. Systems of Linear and Nonlinear Equations

Theorem 12.1.3 For a system of linear equations with two unknowns, if the two equations have the same slope but different intercepts, the system is inconsistent and has no solution. In such cases, the equalization method will lead to a contradiction.

Demostración. Consider a system of two linear equations in the general form:

$$y = m_1 x + b_1$$
$$y = m_2 x + b_2$$

where m_1 and m_2 are the slopes, and b_1 and b_2 are the intercepts.
If $m_1 = m_2$ but $b_1 \neq b_2$, then both equations represent parallel lines with the same slope but different y-intercepts. As parallel lines, they never intersect, meaning there is no point (x, y) that satisfies both equations simultaneously.
Using the equalization method, setting the expressions for y equal:

$$m_1 x + b_1 = m_2 x + b_2.$$

Since $m_1 = m_2$, this simplifies to:

$$b_1 = b_2,$$

which is a contradiction because $b_1 \neq b_2$ by hypothesis. Therefore, the system is inconsistent and has no solution. ∎

Corollary 12.1.4 If applying the equalization method results in a true identity such as $0 = 0$, the system is dependent and has infinitely many solutions. This occurs when the two equations represent the same line.

(R) The equalization method is particularly useful when it is easy to isolate a variable in both equations and one wants to avoid working with fractions initially. However, for systems with more equations or unknowns, other methods, such as elimination or matrix methods, may be more efficient.

Exercise 12.3 Use the equalization method to solve the following system of equations:

$$\begin{cases} 4x + y = 11 \\ x - 2y = -1 \end{cases}$$

Exercise 12.4 Solve the following system using the equalization method:

$$\begin{cases} 3x - y = 4 \\ 2x + y = 5 \end{cases}$$

Verify the solution by substituting the obtained values into both equations.

The resolution of systems through the equalization method simplifies the process of solving simultaneous equations by equating expressions to reduce the number of unknowns, making it easier to solve.

12.1.3 Elimination and System Reduction

Definition 12.1.3 The **elimination method** is a technique for solving systems of linear equations by combining the equations in such a way that one of the unknowns is eliminated. This is achieved by adding or subtracting the equations, often after multiplying them by a suitable value. By eliminating one unknown, the system is reduced to a single equation with one less unknown, simplifying its resolution.

■ **Example 12.3** Consider the system of equations:

$$\begin{cases} 2x + 3y = 13 \\ 4x - y = 5 \end{cases}$$

Applying the elimination method:
1. Multiply the second equation by 3 to make the terms involving y opposites:

$$\begin{cases} 2x + 3y = 13 \\ 12x - 3y = 15 \end{cases}$$

2. Add both equations to eliminate y:

$$2x + 12x + 3y - 3y = 13 + 15 \Rightarrow 14x = 28$$

3. Solve for x:

$$x = 2$$

4. Substitute the value of x into one of the original equations to find y:

$$2(2) + 3y = 13 \Rightarrow 4 + 3y = 13 \Rightarrow 3y = 9 \Rightarrow y = 3$$

Thus, the solution to the system is $x = 2$, $y = 3$. ■

Lema 12.1.3 The elimination method can always be applied if the equations can be multiplied appropriately to obtain opposite terms for the same unknown. If it is not possible to obtain opposite terms through multiplication, the system may be dependent or have no solution.

Theorem 12.1.5 If applying the elimination method leads to an equation that is always true, such as $0 = 0$, then the system has infinitely many solutions and is dependent. If it leads to a contradiction, such as $0 = 5$, then the system has no solution and is inconsistent.

Demostración. When using the elimination method on a system of linear equations, we reduce the system by eliminating variables and simplifying the equations.
1. **Case of an always true equation ($0 = 0$):** If, at the end of the elimination process, we obtain an equation like $0 = 0$, this indicates that one of the equations is a linear combination of the others, implying dependence among the equations. This means the system has infinitely many solutions, as there is a dependency and multiple values of the variables can satisfy the equations.
2. **Case of a contradiction ($0 = 5$):** If we arrive at an equation like $0 = 5$, which is a contradiction, it means there are no values of the variables that can simultaneously satisfy all the equations. Therefore, the system is inconsistent and has no solution.
Thus, the nature of the system depends on whether the elimination process yields a trivial equation or a contradiction. ∎

Corollary 12.1.6 For any consistent system of linear equations, the elimination method produces a unique solution if the number of equations equals the number of unknowns and all the equations are linearly independent.

Demostración. If the system is consistent, it has at least one solution. When the number of equations equals the number of unknowns and all the equations are linearly independent, the coefficient matrix of the system has full rank, meaning its rank equals the number of unknowns.
Using the elimination method, the system is reduced to a form where each equation provides unique information about one variable. This ensures that the system has no redundant equations and that each variable can be uniquely determined. Thus, the system has a unique solution under these conditions. ∎

(R) The elimination method is particularly useful when the coefficients of the unknowns can be easily manipulated to obtain opposite terms. However, in some cases, the substitution method or the use of matrices may be more efficient depending on the structure of the system.

Exercise 12.5 Use the elimination method to solve the following system of equations:

$$\begin{cases} 3x - 2y = 7 \\ 2x + 4y = -2 \end{cases}$$

Exercise 12.6 Solve the following system using the elimination method:

$$\begin{cases} 5x + 6y = 16 \\ -10x + 9y = -7 \end{cases}$$

Verify your solution by substituting the obtained values into both equations.

Elimination and system reduction simplify solving complex systems by combining and eliminating unknowns, making it easier to obtain results efficiently and systematically.

12.2 Application of the Gauss-Jordan Method

12.2.1 Solving Systems Using the Gauss-Jordan Method

Definition 12.2.1 The **Gauss-Jordan method** is a technique for solving systems of linear equations by transforming the augmented matrix of the system into its row-reduced echelon form. The goal is to obtain an identity matrix in the coefficient part, which directly provides the solutions to the system. This method is an extension of Gaussian elimination.

■ **Example 12.4** Let us solve the following system of linear equations using the Gauss-Jordan method:

$$\begin{cases} x + y + z = 6 \\ 2x + 3y + z = 10 \\ x + 2y + 3z = 13 \end{cases}$$

12.2 Application of the Gauss-Jordan Method

First, represent the system as its augmented matrix:

$$\begin{bmatrix} 1 & 1 & 1 & | & 6 \\ 2 & 3 & 1 & | & 10 \\ 1 & 2 & 3 & | & 13 \end{bmatrix}$$

Then, perform row operations to reduce the matrix to its row-reduced echelon form:
1. Subtract $2\times$ (R1) from (R2) to eliminate the x-coefficient in the second row:

$$\begin{bmatrix} 1 & 1 & 1 & | & 6 \\ 0 & 1 & -1 & | & -2 \\ 1 & 2 & 3 & | & 13 \end{bmatrix}$$

2. Subtract (R1) from (R3):

$$\begin{bmatrix} 1 & 1 & 1 & | & 6 \\ 0 & 1 & -1 & | & -2 \\ 0 & 1 & 2 & | & 7 \end{bmatrix}$$

3. Subtract (R2) from (R3):

$$\begin{bmatrix} 1 & 1 & 1 & | & 6 \\ 0 & 1 & -1 & | & -2 \\ 0 & 0 & 3 & | & 9 \end{bmatrix}$$

4. Divide (R3) by 3:

$$\begin{bmatrix} 1 & 1 & 1 & | & 6 \\ 0 & 1 & -1 & | & -2 \\ 0 & 0 & 1 & | & 3 \end{bmatrix}$$

5. Add (R3) to (R2):

$$\begin{bmatrix} 1 & 1 & 1 & | & 6 \\ 0 & 1 & 0 & | & 1 \\ 0 & 0 & 1 & | & 3 \end{bmatrix}$$

6. Subtract (R3) from (R1):

$$\begin{bmatrix} 1 & 1 & 0 & | & 3 \\ 0 & 1 & 0 & | & 1 \\ 0 & 0 & 1 & | & 3 \end{bmatrix}$$

Finally, subtract (R2) from (R1):

$$\begin{bmatrix} 1 & 0 & 0 & | & 2 \\ 0 & 1 & 0 & | & 1 \\ 0 & 0 & 1 & | & 3 \end{bmatrix}$$

Thus, the solution to the system is $x = 2$, $y = 1$, $z = 3$. ∎

Lema 12.2.1 The Gauss-Jordan method can always be applied to a consistent system of linear equations. If the augmented matrix can be fully reduced to row-reduced echelon form, the system has a unique solution.

Theorem 12.2.1 If the augmented matrix of a linear system can be reduced using the Gauss-Jordan method to a matrix where there is at least one row with all elements equal to zero (both in the coefficient part and in the augmented column), then the system is dependent and has infinitely many solutions. If there is a row of zeros in the coefficient part and a nonzero value in the augmented column, then the system is inconsistent and has no solution.

Corollary 12.2.2 A linear system has a unique solution if the augmented matrix, when reduced by the Gauss-Jordan method, produces an identity matrix in the coefficient part. This implies that the system's equations are linearly independent.

(R) The Gauss-Jordan method is an extension of Gaussian elimination and is particularly useful when we want to obtain the solution to a system directly in one step, especially when working with large matrices.

Exercise 12.7 Solve the following system of equations using the Gauss-Jordan method:

$$\begin{cases} x + 2y + 3z = 9 \\ 2x + y + z = 8 \\ 3x + 2y + z = 10 \end{cases}$$

Exercise 12.8 Use the Gauss-Jordan method to solve the following system and determine if it has a unique solution, infinitely many solutions, or if it is inconsistent:

$$\begin{cases} 2x + y - z = 1 \\ 4x - y + 2z = 2 \\ -2x + 3y + 5z = -4 \end{cases}$$

The Gauss-Jordan method, by reducing systems to their simplest form, is fundamental for efficiently solving linear systems, and its connection to matrix inversion and consistency determination is key in linear algebra.

12.2.2 Augmented Matrices and Row Reduction

Definition 12.2.2 An **augmented matrix** is a representation of a system of linear equations in matrix form, where an additional column is included to represent the constant terms of each equation. For example, given a system of two equations with two unknowns:

12.2 Application of the Gauss-Jordan Method

$$\begin{cases} a_1x + b_1y = c_1 \\ a_2x + b_2y = c_2 \end{cases}$$

The augmented matrix of the system is:

$$\left[\begin{array}{cc|c} a_1 & b_1 & c_1 \\ a_2 & b_2 & c_2 \end{array}\right]$$

The use of the augmented matrix allows the application of techniques such as row reduction to solve the system.

■ **Example 12.5** Consider the following system of linear equations:

$$\begin{cases} 2x + 3y = 8 \\ 4x - y = 2 \end{cases}$$

The corresponding augmented matrix is:

$$\left[\begin{array}{cc|c} 2 & 3 & 8 \\ 4 & -1 & 2 \end{array}\right]$$

We can apply the row reduction method to simplify the matrix into row-reduced echelon form and solve the system:
1. Subtract $2\times$ row 1 from row 2:

$$\left[\begin{array}{cc|c} 2 & 3 & 8 \\ 0 & -7 & -14 \end{array}\right]$$

2. Divide row 2 by -7:

$$\left[\begin{array}{cc|c} 2 & 3 & 8 \\ 0 & 1 & 2 \end{array}\right]$$

3. Subtract $3\times$ row 2 from row 1:

$$\left[\begin{array}{cc|c} 2 & 0 & 2 \\ 0 & 1 & 2 \end{array}\right]$$

Finally, divide row 1 by 2:

$$\left[\begin{array}{cc|c} 1 & 0 & 1 \\ 0 & 1 & 2 \end{array}\right]$$

The solution to the system is $x = 1, y = 2$. ■

Lema 12.2.2 If a system of linear equations has an augmented matrix that, through row reduction, results in a matrix where a row of zeros is followed by a nonzero value in the column of constants, then the system has no solution, i.e., it is inconsistent.

> **Theorem 12.2.3** If the augmented matrix of a system of linear equations can be reduced to a matrix in row-reduced echelon form where the coefficient part forms an identity matrix, then the system has a unique solution and is consistent.

Demostración. Suppose the augmented matrix of the system can be reduced using elementary row operations to row-reduced echelon form, in which the submatrix of coefficients becomes the identity matrix. This implies that the original system has the same number of equations as unknowns and that the equations are linearly independent.

In this form, each equation corresponds to a variable taking a specific value, as each row of the identity matrix fixes the value of a variable without redundancy. This ensures that the system has a unique solution for each variable.

Since a solution has been reached without contradictions, the system is consistent. Thus, if the augmented matrix is reduced to this form, the system has a unique solution and is consistent. ∎

> **Corollary 12.2.4** If a system of linear equations has an augmented matrix that can be reduced to a matrix where all nonzero rows have pivots (i.e., nonzero leading entries), then the system has a unique solution. This occurs because the number of linearly independent equations equals the number of unknowns.

Demostración. If the augmented matrix of a system of linear equations is reduced to a form where each nonzero row has a pivot, then the coefficient submatrix has full rank equal to the number of unknowns. This implies that the equations are linearly independent, and each variable can be uniquely solved.

The presence of a pivot in each nonzero row ensures that the system does not contain redundant or inconsistent equations. Since the number of independent equations equals the number of unknowns, the system has a unique solution.

Therefore, the existence of pivots in all nonzero rows guarantees that the system is consistent and has a unique solution. ∎

> (R) Row reduction is a fundamental technique for solving systems of linear equations and allows for straightforward identification of the system's consistency, the existence of solutions, and whether those solutions are unique or infinite.

> **Exercise 12.9** Solve the following system of equations using row reduction:
>
> $$\begin{cases} 3x + 2y - z = 7 \\ 2x - 2y + 4z = 10 \\ -x + \frac{1}{2}y - z = -1 \end{cases}$$

> **Exercise 12.10** Determine whether the following system of equations has a unique solution, infinite solutions, or is inconsistent using the row reduction method:

12.2 Application of the Gauss-Jordan Method

$$\begin{cases} x+y+z=3 \\ 2x+2y+2z=6 \\ 3x+3y+3z=9 \end{cases}$$

These definitions, theorems, and exercises provide a comprehensive view of using augmented matrices and the row reduction process to solve linear systems, connecting concepts like consistency, uniqueness of solutions, and fundamental algebraic techniques.

12.2.3 Practical Examples of Solutions

■ **Example 12.6 Example 1: 2x2 Matrix**

Let us solve the following system of linear equations:

$$\begin{cases} 2x+3y=8 \\ 4x-y=2 \end{cases}$$

The augmented matrix of the system is:

$$\begin{bmatrix} 2 & 3 & | & 8 \\ 4 & -1 & | & 2 \end{bmatrix}$$

We apply the row reduction method:
1. Subtract $2\times$ row 1 from row 2:

$$\begin{bmatrix} 2 & 3 & | & 8 \\ 0 & -7 & | & -14 \end{bmatrix}$$

2. Divide row 2 by -7:

$$\begin{bmatrix} 2 & 3 & | & 8 \\ 0 & 1 & | & 2 \end{bmatrix}$$

3. Subtract $3\times$ row 2 from row 1:

$$\begin{bmatrix} 2 & 0 & | & 2 \\ 0 & 1 & | & 2 \end{bmatrix}$$

Finally, divide row 1 by 2:

$$\begin{bmatrix} 1 & 0 & | & 1 \\ 0 & 1 & | & 2 \end{bmatrix}$$

The solution to the system is $x=1, y=2$. ■

■ Example 12.7 Example 2: 3x3 Matrix

Consider the following system of equations:

$$\begin{cases} x + 2y + z = 9 \\ 2x - y + 3z = 8 \\ 3x + y + 2z = 10 \end{cases}$$

The corresponding augmented matrix is:

$$\begin{bmatrix} 1 & 2 & 1 & | & 9 \\ 2 & -1 & 3 & | & 8 \\ 3 & 1 & 2 & | & 10 \end{bmatrix}$$

We apply row reduction:
1. Subtract $2\times$ row 1 from row 2 and $3\times$ row 1 from row 3:

$$\begin{bmatrix} 1 & 2 & 1 & | & 9 \\ 0 & -5 & 1 & | & -10 \\ 0 & -5 & -1 & | & -17 \end{bmatrix}$$

2. Subtract row 2 from row 3:

$$\begin{bmatrix} 1 & 2 & 1 & | & 9 \\ 0 & -5 & 1 & | & -10 \\ 0 & 0 & -2 & | & -7 \end{bmatrix}$$

3. Divide row 3 by -2:

$$\begin{bmatrix} 1 & 2 & 1 & | & 9 \\ 0 & -5 & 1 & | & -10 \\ 0 & 0 & 1 & | & \frac{7}{2} \end{bmatrix}$$

Finally, we obtain the value of each variable: $z = \frac{7}{2}, y = -1, x = 2$. ■

■ Example 12.8 Example 3: 3x3 Matrix

Consider the following system:

$$\begin{cases} x - y + z = 2 \\ 2x + y - 3z = -3 \\ -3x + 2y + 2z = 5 \end{cases}$$

The augmented matrix is:

$$\begin{bmatrix} 1 & -1 & 1 & | & 2 \\ 2 & 1 & -3 & | & -3 \\ -3 & 2 & 2 & | & 5 \end{bmatrix}$$

By applying row reduction and simplifying, we find the solution $x = 1, y = 0, z = 3$. ■

12.3 Nonlinear Systems of Equations and Their Applications

■ Example 12.9 **Example 4: 4x4 Matrix**
Consider a system of four equations with four unknowns:

$$\begin{cases} x+y+z+w=4 \\ 2x-y+3z+2w=5 \\ -x+4y+2z-w=1 \\ 3x+y-z+4w=10 \end{cases}$$

The augmented matrix is:

$$\begin{bmatrix} 1 & 1 & 1 & 1 & | & 4 \\ 2 & -1 & 3 & 2 & | & 5 \\ -1 & 4 & 2 & -1 & | & 1 \\ 3 & 1 & -1 & 4 & | & 10 \end{bmatrix}$$

We apply row reduction to simplify the matrix and find the solution. ■

■ Example 12.10 **Example 5: 4x4 Matrix**
Consider the following system:

$$\begin{cases} x+2y+3z+4w=10 \\ 2x+3y+z+5w=12 \\ 3x-y+4z+w=8 \\ 4x+y+2z+3w=15 \end{cases}$$

The augmented matrix is:

$$\begin{bmatrix} 1 & 2 & 3 & 4 & | & 10 \\ 2 & 3 & 1 & 5 & | & 12 \\ 3 & -1 & 4 & 1 & | & 8 \\ 4 & 1 & 2 & 3 & | & 15 \end{bmatrix}$$

We apply row reduction to simplify the matrix and obtain the solution to the system. ■

These examples cover matrices of orders 2x2, 3x3, and 4x4, demonstrating the application of row reduction techniques to solve systems of equations effectively.

12.3 Nonlinear Systems of Equations and Their Applications

12.3.1 Definition of Nonlinear Systems

Definition 12.3.1 A **nonlinear system of equations** is a set of two or more equations involving multiple variables, where at least one of the equations is nonlinear. Nonlinear equations include terms that are not of the first degree, such as squares, roots, products of variables, trigonometric functions, exponentials, logarithms, and others.

■ Example 12.11 Consider the following system:

$$\begin{cases} x^2+y^2=25 \\ x+y=7 \end{cases}$$

To solve it, we can isolate one of the variables in the linear equation, for example, $y = 7 - x$, and substitute it into the quadratic equation:

$$x^2 + (7-x)^2 = 25$$

Expanding and solving, we find the possible values of x and y. ∎

Lema 12.3.1 If a nonlinear system includes an equation of the form $f(x,y) = g(x,y)$, where both functions are continuous, the system can be solved using the substitution method if one of the variables can be conveniently isolated.

> Theorem 12.3.1 **Existence and Uniqueness Theorem for Nonlinear Systems**
> If a nonlinear system of equations has a unique solution in an open region of the plane, the solution will be continuous with respect to the system's parameters. In other words, small variations in the system's coefficients lead to small variations in the solution.

Demostración. For a nonlinear system of equations, the existence of a unique solution in an open region of the plane implies that the conditions for continuity and differentiability are satisfied in the vicinity of the solution. The continuity of the solution with respect to the parameters follows from the theorems of continuity and differentiability for multivariable functions.
Specifically, as the coefficients of the system vary slightly, the functions describing the equations also change continuously within that region. Since the solution is unique, any small variation in the parameters results in only a small variation in the solution, preserving its uniqueness and continuity relative to the system's parameters.
Thus, the solution to the nonlinear system is continuous with respect to the system's parameters within the considered region. ∎

> **Corollary 12.3.2** If a nonlinear system of equations has a unique solution, the graphs of the corresponding functions will intersect at a single point. This observation enables the use of graphical techniques to determine the existence of solutions.

 When solving a nonlinear system, it is helpful to analyze whether substitution or elimination methods can simplify the solution process. Additionally, graphical analysis can be a powerful tool for visualizing the existence of intersection points.

Exercise 12.11 Solve the following nonlinear system:

$$\begin{cases} x^2 + y^2 = 10 \\ x - y = 2 \end{cases}$$

Use the substitution method to find the possible values of x and y. ∎

Exercise 12.12 Determine whether the following system has a solution, and if so, find it:

$$\begin{cases} e^x + y = 4 \\ \sin(y) = x \end{cases}$$

12.3 Nonlinear Systems of Equations and Their Applications

> Remember that you can use numerical approximations to solve equations involving transcendental functions.

Each of these results is interrelated to provide a clearer understanding of how to approach the resolution of nonlinear systems of equations. For example, the existence and uniqueness theorem helps determine when a system may have a unique solution, which is fundamental when applying methods like substitution.

12.3.2 Methods for Solving Nonlinear Systems

Definition 12.3.2 The **methods for solving nonlinear systems of equations** include various techniques used depending on the system's complexity and the type of equations involved. Common methods include the substitution method, the elimination method, the equalization method, and numerical techniques such as the Newton-Raphson method.

■ **Example 12.12** Consider the following nonlinear system:

$$\begin{cases} x^2 + y^2 = 13 \\ y = x + 1 \end{cases}$$

To solve it, we use the substitution method. First, substitute $y = x + 1$ into the first equation:

$$x^2 + (x+1)^2 = 13$$

Expand and solve the resulting quadratic equation to find the possible values of x and subsequently y. ■

Lema 12.3.2 The elimination method can be used to solve nonlinear systems as long as at least one of the equations can be manipulated to eliminate a variable conveniently. This method is generally more effective when both equations are of similar degrees.

> Theorem 12.3.3 **Continuity Theorem for Solutions**
> If a nonlinear system is composed of continuous and differentiable functions in a region, the system has continuous solutions in the same region. Furthermore, if the system has multiple solutions, these solutions will also be differentiable functions.

Corollary 12.3.4 If a nonlinear system has a solution in a bounded region and both equations are differentiable functions, the intersection of their curves can be found using numerical methods such as the Newton-Raphson method, iterating to achieve greater precision.

(R) The use of numerical methods like Newton-Raphson is particularly useful for solving more complex nonlinear systems. This method relies on derivatives and convergence, so it is important to verify initial conditions and ensure the existence of continuous derivatives in the region of interest.

Exercise 12.13 Solve the following nonlinear system using the substitution method:

$$\begin{cases} y = x^2 + 3 \\ y + x = 6 \end{cases}$$

Find all possible values of x and y.

Exercise 12.14 Use the equalization method to solve the system:

$$\begin{cases} x^2 + y^2 = 20 \\ x - y = 2 \end{cases}$$

Determine whether the system has a unique solution or multiple solutions.

Each of the mentioned methods, such as substitution and equalization, has its advantages and limitations depending on the system's nature. In some cases, it is convenient to combine methods or use numerical techniques to approximate solutions that cannot be found exactly.

12.3.3 Applications of Nonlinear Systems in Real-World Problems

Definition 12.3.3 A **nonlinear system** is a set of equations where at least one equation is nonlinear, meaning it may include terms such as powers, roots, products of variables, or nonlinear functions like sine or logarithms. These systems are useful for modeling complex phenomena that do not follow proportional relationships.

■ **Example 12.13** Suppose we want to model the growth of two interacting populations, P_1 and P_2, such as in the predator-prey model:

$$\begin{cases} \frac{dP_1}{dt} = r_1 P_1 - a P_1 P_2 \\ \frac{dP_2}{dt} = -r_2 P_2 + b P_1 P_2 \end{cases}$$

Here, r_1, r_2, a, b are constants representing growth rates, predation, and reproduction. This nonlinear system exemplifies how biological interactions can be represented using nonlinear differential equations. ■

Lema 12.3.3 In a nonlinear system, the existence of a solution depends on the continuity and differentiability of the functions involved. If the functions are continuous and behave smoothly, the system is more likely to have solutions within the considered range.

Theorem 12.3.5 Local Stability Theorem

If a nonlinear system has a solution (x_0, y_0), and both functions defining the system are differentiable around this point, the stability of the solution can be analyzed using the eigenvalues of the Jacobian matrix evaluated at that point. This theorem is commonly applied in systems modeling physical or biological phenomena.

Corollary 12.3.6 If a nonlinear system has a critical point and the Jacobian matrix evaluated at this point has all its eigenvalues negative, then the critical point is a stable equilibrium.

> Nonlinear systems are essential for modeling complex real-world phenomena such as species interaction, fluid dynamics, and economics. Their analysis often requires advanced techniques, including numerical methods or linearization around critical points.

Exercise 12.15 Consider an electric circuit with a nonlinear resistor described by:

$$\begin{cases} V = IR(I) \\ I = 3 - e^{-\frac{V}{2}} \end{cases}$$

Solve the system to find the current I as a function of the voltage V. Physically interpret the obtained solution.

Exercise 12.16 A water tank has an inflow and an outflow that depend nonlinearly on the water level $h(t)$. The system is described as follows:

$$\begin{cases} \frac{dh}{dt} = k_1 h - k_2 \sqrt{h} \\ h(0) = h_0 \end{cases}$$

Determine if there is an equilibrium point, and if it exists, analyze whether it is stable or unstable.

The examples above illustrate how nonlinear systems frequently arise in real-world problems, from circuit analysis to fluid dynamics. Solving these systems requires advanced mathematical skills that combine algebraic methods and numerical techniques to find solutions, enabling a deeper understanding of the system's behavior.

12.4 Solved Exercises

Exercise 12.17 Solve the following system of linear equations using the substitution method:

$$\begin{cases} x + 2y = 8 \\ 3x - y = 5 \end{cases}$$

Demostración. From the first equation, solve for x:

$$x = 8 - 2y$$

Substitute into the second equation:

$$3(8 - 2y) - y = 5$$

$$24 - 6y - y = 5 \implies -7y = -19 \implies y = \frac{19}{7}$$

Substitute y into $x = 8 - 2y$:

$$x = 8 - 2\left(\frac{19}{7}\right) = \frac{56 - 38}{7} = \frac{18}{7}$$

Thus, $x = \frac{18}{7}, y = \frac{19}{7}$. ∎

Exercise 12.18 Solve the nonlinear system:

$$\begin{cases} x^2 + y^2 = 25 \\ y = x + 3 \end{cases}$$

Demostración. Substitute $y = x + 3$ into the first equation:

$$x^2 + (x+3)^2 = 25$$

$$x^2 + x^2 + 6x + 9 = 25$$

$$2x^2 + 6x - 16 = 0 \implies x^2 + 3x - 8 = 0$$

Solve the quadratic equation:

$$x = \frac{-3 \pm \sqrt{9 + 32}}{2} = \frac{-3 \pm 7}{2}$$

$$x_1 = 2, \quad x_2 = -5$$

For $x = 2$, $y = 5$, and for $x = -5$, $y = -2$. The solutions are $(2, 5)$ and $(-5, -2)$. ■

12.5 Proposed Exercises

12.5.1 Methods for Solving Linear Systems

Exercise 12.19 Solve the system of equations $x + y = 4$, $2x - y = 1$ using the substitution method.

Exercise 12.20 Determine the solution of the system $3x + 2y = 8$, $x - y = 1$ using the elimination method.

Exercise 12.21 Apply the elimination method to solve the system $x + 3y = 9$, $4x - y = 5$.

Exercise 12.22 Use the reduction method to solve the system $2x + 4y = 10$, $-x + y = 3$.

Exercise 12.23 Solve the linear system using the substitution method: $x - y = 2$, $3x + y = 11$.

12.5.2 Application of Gauss-Jordan Method

Exercise 12.24 Solve the following system using the Gauss-Jordan method:

$$\begin{cases} x + y + z = 6 \\ 2x - y + 3z = 14 \\ -x + 4y - z = -2 \end{cases}$$

Exercise 12.25 Use the Gauss-Jordan method to solve the system:

$$\begin{cases} 2x - 3y + z = 5 \\ 4x + y - 2z = -2 \\ -3x + 2y + 4z = 3 \end{cases}$$

Exercise 12.26 Solve the system using the Gauss-Jordan method:

$$\begin{cases} x + y + 2z = 7 \\ 2x - y + z = 3 \\ 3x + y - z = 4 \end{cases}$$

12.5.3 Nonlinear Systems and Applications

Exercise 12.27 Solve the nonlinear system:

$$\begin{cases} x^2 + y^2 = 25 \\ y = x + 3 \end{cases}$$

Exercise 12.28 Find the points of intersection between the parabola $y = x^2$ and the line $y = 2x + 3$.

Exercise 12.29 Solve the system of equations:

$$\begin{cases} y = x^3 \\ x + y = 4 \end{cases}$$

Exercise 12.30 Determine the solution for the nonlinear system:

$$\begin{cases} x^2 + y = 6 \\ y^2 = 4x \end{cases}$$

Exercise 12.31 Find the solutions of the system:

$$\begin{cases} x^2 + y^2 = 9 \\ y = 2x - 1 \end{cases}$$

Índice alfabético

A

Absolute Value	25
Absolute value and composition	179
Absolute value inequality	47
Addition	27
addition of complex numbers	65
Addition of functions	149
additive identity element	66
Additive identity of functions	153
adjoint of a matrix	230
Applications of conics	141
Applications of inverse functions	195
Applications of the circle	131
Applications of the hyperbola	140
Applications of the hyperbola in GPS	141
Applications of the parabola	136
arccos	215
Area of the circle	131
Argand plane	59
Argument of a complex number	62
Associative property	20
Associative property of functions	153
associativity	220, 222
Associativity of composition	177, 183, 185
Augmented Matrix	251

B

bacterial growth	211
Bijective function	187
Bijectivity and inverse	191
Bijectivity of composition	177
Bounds and scaling	171

C

Call cost	158
Cartesian equation of a line	97
Center and radius theorem	130
Center of the circle	127, 129
Circle	127
Circle as a special case of the ellipse	138
Circle reduced to a point	130
Circular gear	131
Closure	17
Closure corollary	18
Closure exercise	18
Closure property	17
Closure theorem	17
Closure under addition	149
cofactors and adjoint	230
coincident lines	97
Commutative property	20

Commutative property of functions 152
commutativity 220
commutativity of complex multiplication .. 67
Completing the square 129, 130
complex conjugate 68, 69
Composite function 146
Composite functions 147, 148
Composition of quadratic and root functions 178
Composition of transformations 168
Composition of trigonometric functions .. 185
Condition for parallelism of planes 120
Condition for perpendicularity of planes .. 119
Condition for the existence of a circle 129
Conjugate axis 139
Conjugate of a complex number 69
Consistency of Linear Systems 244
Continuity and scaling 175
Continuity in real-world models 158
Continuity of a function 148
Continuity of composite functions 178
Continuity of function division 151
Continuity of function multiplication 151
Continuity of piecewise functions ... 155, 157
continuity of solutions in nonlinear systems 257
Continuity of the sum of functions 149
Continuity of transformed functions 168
Continuity under horizontal scaling 174
Continuity under translation 164
Continuous derivative of piecewise functions 155, 157
Conversion between forms of the equation 110
Conversion to parametric form 111
Corollary for graphical representation in two variables 50
Corollary for homogeneous equations 38
Corollary of boundedness in absolute value inequalities 48
Corollary of commutative and associative properties 22
Corollary of complex roots 40
Corollary of conjugate roots 41
Corollary of distributive property 20
Corollary of inequalities with positive terms 49
corollary of inverse matrices 233
Corollary of nested radicals 43
Corollary on absolute value inequality 26
Corollary on conjugate in the Argand plane 61
Corollary on conjugates of complex numbers 58
Corollary on division reciprocity 30
Corollary on interval endpoints 25
Corollary on interval inclusion 23
Corollary on multiple radical elimination .. 46
Corollary on multiplication in polar form .. 64
Corollary on Negative Powers of Complex Numbers 75
Corollary on power of a root 32
Corollary on rationalization with conjugates 45
Corollary on symmetric subtraction 29
Corollary on the conjugate of a sum 71
Corollary on the distribution of roots in the complex plane 73
Corollary on the existence of the inverse for monotonic functions 190
Corollary on the symmetry of nth roots 78
Corollary, Consistent Systems 248
Corollary, Dependent or Inconsistent Systems 244
Corollary, Dependent Systems 246
corollary, determinant and inverse matrix . 237
Corollary, Unique Solution 252
Corollary, Unique Solution of Systems ... 250
Criterion for the existence of an inverse function 189
Criterion for verifying the inverse 191
cross product 94
cross product of parallel vectors 94

D

De Moivre's Formula 72, 74
De Moivre's Theorem 67, 69
Definition of piecewise functions 154
derivative of exponential function 204
derivative of exponential growth 211
Derivative of function products 152
Derivative of function quotients 152
Derivative of polynomial functions 150
Derivative of the difference of functions .. 150
Derivative of the sum of functions 149
determinant 226, 228
determinant and multiplication 229
determinant example 227
determinant of matrix with proportional rows 229
determinant of transpose 225
determinant of triangular matrix 227

determinant of zero row 229
determinant properties 228, 229, 231
Differentiability of functions 151
Differentiability of piecewise functions .. 155, 157
Directrix 132
Directrix of the parabola 134
Distance between parallel lines 114
Distance from a point to a line 107
Distributive Property 19
Distributive property of functions 152
Division 29
division of complex numbers 68
Division of functions 151
Domain 145, 148
Domain and range of composite functions 148
Domain of a function 148
Domain of composite functions 146, 148
Domain of radical functions 147
Domain of rational functions 147
Domain of rational functions with quadratic denominator 147
Dot product 121
dot product 84, 90
Dot product and perpendicularity 112
Downward scaling 170

E

Eccentricity 138
Eccentricity calculation 140
Eccentricity comparison 140
Eccentricity of the ellipse 137
Eccentricity of the hyperbola 140
elimination in nonlinear systems 257
Elimination Method 247
Elimination Method, Applicability 247
Elimination Method, Matrix Method 246
Ellipse 137, 141
equality of vectors 84
Equalization Method 245
Equalization Method, Applicability 245
Equation equivalence 109
equation of a plane 99
Equation of the circle 127
Equation of the ellipse 137
Equation of the hyperbola 139
Equation of the parabola 134
Equilateral triangle 116

Equivalence of parametric and vector forms 110
Even function 165
Example of a first-degree equation 37
Example of a nonlinear inverse function .. 187
example of a plane equation 99
Example of absolute value 25
Example of absolute value in composition 185
Example of absolute value inequality 47
Example of addition and subtraction 28
Example of an inverse function 186
Example of closure 17
Example of commutative and associative properties 21
Example of composition 176, 183
Example of composition notation 184
Example of composition with roots and powers 186
Example of cube roots of a complex number 76
Example of cubic graph and inverse 195
Example of distributive property 19
Example of exponential composition 186
example of exponential equation 208
Example of factoring a higher-degree equation 41
Example of factoring a second-degree equation 39
Example of function reflection 165
Example of function translation 163
Example of graph of a composite function 178
Example of graph of an inverse function .. 195
Example of graphical representation 50
Example of graphical transformations 167
Example of graphing piecewise functions 156
Example of horizontal scaling 172
Example of interval 22, 24
Example of inverse function in finance ... 196
example of inverse functions 206
Example of inverse of a quadratic function with restricted domain 190
example of line equation 97
Example of linear translation 164
example of logarithm 205
Example of logarithmic composition 186
example of logarithmic equation 209
Example of modulus and argument 63
Example of multiplication and division ... 29
Example of piecewise function 154
Example of polynomial function composition 185

Example of powers and roots 31
Example of Powers of Complex Numbers . 74
Example of radicals 42
Example of representation in the Argand plane 60
Example of scaling 170
Example of scaling in graphs 175
Example of simplifying and rationalizing radicals 44
Example of sinusoidal composition 185
Example of solving a radical equation 46
Example of solving inequalities 49
Example of symmetry of inverse function 194
Example of the conjugate of a complex number 69
Example of the method for finding the inverse function 189
example of transpose 224
example of vector addition and subtraction 88
Example of verifying inverse functions . . . 192
Example of verifying the inverse function 191
Example using De Moivre's Formula 72
Example, Augmented Matrix 251
Example, Elimination Method 247
Example, Equalization Method 245
Example, Gauss-Jordan Method 249
Exercise on absolute value 27
Exercise on absolute value inequalities 48
Exercise on addition and subtraction 29
Exercise on commutative and associative properties 22
Exercise on De Moivre's Formula 74
Exercise on distributive property 20
Exercise on exponential inverse function . 196
Exercise on first-degree equations 38
Exercise on graphical representation of solutions 51
Exercise on higher-degree equations 42
Exercise on intervals 23
Exercise on modulus and argument of complex numbers 65
Exercise on multiplication and division 31
Exercise on nth roots of complex numbers . 78
Exercise on operations with complex numbers 59
Exercise on powers and roots 33
Exercise on Powers of Complex Numbers Using De Moivre's Formula 76
Exercise on radical equations 47
Exercise on representation in the Argand plane 62
Exercise on second-degree equations 40
Exercise on simplification and rationalization of radicals 46
Exercise on simplifying radicals 44
Exercise on solving inequalities 49
Exercise on the conjugate of a complex number 71
Exercise on types of intervals 25
Existence and Uniqueness Theorem 244
existence and uniqueness theorem 256
existence of nonlinear solutions 258
exponential decay 210
exponential equation 208
exponential function 203, 206
exponential growth 203, 206, 210
exponential growth and decay theorem . . . 211
exponential growth theorem 204

F

First-degree equation 37
Focal distance 137, 138
Focal distance of the hyperbola 139
Foci of the ellipse 137
Foci of the hyperbola 139
Focus 132
Focus and directrix 133, 135
Focus of the parabola 134
Frequency and horizontal scaling 173
Function 145
Function composition 176, 183, 184
Function domain 145
Function range 145
fundamental relationships 213
Fundamental Theorem of Algebra 41

G

Gauss-Jordan Method 248
Gauss-Jordan method 234
Gauss-Jordan method vs numerical methods 235
Gauss-Jordan Method, Consistency 250
Gauss-Jordan theorem 235
Gaussian Elimination Method 244
Gaussian Elimination, Extension 250

ÍNDICE ALFABÉTICO

Gaussian elimination, LU factorization ... 237
General equation of a line 105
General equation of the circle 128
General equation of the parabola 133, 135
General form of the plane equation 117
Geometric problems 115
Graph of a sinusoidal composition 179
Graph of an inverse function 195
Graph of composite functions 178
Graph of piecewise functions 156, 157
graph of the logarithmic function 207
graphical inverses 207
Graphical representation 148
Graphical representation of domain and range 148
Graphical representation of solutions 50
Graphical symmetry of inverse functions . 196
Graphical transformation 167
Growth direction of composite functions . 178
Growth of functions under translation 164

H

half-life 211
Higher-degree equation 40
Horizontal scaling 172
Horizontal scaling of trigonometric functions 175
Horizontal translation 163
Hyperbola 139, 141
Hyperbola and escape trajectories 141
Hyperbolic trajectory 141

I

Identity function and composition 184
identity matrix 223
Identity property 154
Inconsistent Systems 251
Inequality 49
Injective function 146
Injectivity and inverse function 189
Injectivity and surjectivity of compositions 184
injectivity of exponential function 204
Intersection of graph of function and inverse 195
Intersection of planes 121
Intersection of three planes 122

Interval 22
    Closed 24
    Half-Open 24
    Open 24
interval of inverse functions 215
Inverse function 186, 192
inverse matrix 223, 232
inverse matrix calculation 232
inverse matrix calculation using Gauss-Jordan 234
inverse matrix, system of equations 236
Inverse of monotonic functions 191
inverse of sine 215
inverse of tangent 215
inverse trigonometric functions 215
Inverses of composed functions 185
invertibility corollary 235
invertibility lemma and Gauss-Jordan method 235
invertibility theorem of a matrix 233
invertible matrix 227
involutive property of transpose 224
Inward scaling 172
Isosceles triangle 115

L

Lemma for graphical representation of inequalities 50
Lemma for solving first-degree equations .. 37
Lemma of absolute value inequalities 47
Lemma of closed interval 22
Lemma of commutative and associative properties 21
Lemma of distributive property 19
Lemma of radicals under multiplication ... 42
Lemma of sign change in inequalities 49
Lemma of the discriminant 39
Lemma of the Factor Theorem 41
Lemma on addition of complex numbers .. 58
Lemma on commutativity and associativity of addition 28
Lemma on commutativity and associativity of multiplication 29
Lemma on intervals 24
Lemma on magnitude of a complex number 60
Lemma on powers 31
Lemma on product of absolute values 26
Lemma on product under radicals 44

Lemma on radical elimination 46
Lemma on the distribution of nth roots 77
Lemma on the Modulus of Powers of Complex Numbers 74
Lemma on the modulus using De Moivre's Formula 72
Lemma on the product of a complex number and its conjugate 69
Lemma on triangular property of modulus . 63
Limited range 148
Line of intersection of planes 122
Linear combination 109
linear independence 95
Linear piecewise models 159
linearity of determinant 229
linearity of transpose 225
logarithm 205
Logarithm and exponential composition .. 179
logarithm-exponential relationship 210
logarithmic equation 209
logarithmic function 206
logarithmic growth 205
logarithmic to exponential conversion 210

M

magnitude of a complex number 67
Mathematical model 158
matrix addition 219, 220
matrix inverse 231
matrix multiplication 222
matrix subtraction 219, 220
Maximum height calculation 136
Maximum height of a projectile 136
maximum value of trigonometric functions 212
Median of a triangle 116
Method for finding the inverse function .. 188
methods for calculating the inverse matrix 233
Modulus of a complex number 62
Monotonic functions and inverses 192
Multiplication 29
multiplication of complex numbers 67
Multiplication of functions 151
Multiplicative identity of functions 153

N

Non-commutative composition 176

non-commutative multiplication 224
Non-commutativity of composition 183
nonlinear systems 258
nonlinear systems of equations 255
nonlinear systems, solving methods 257
norm of a vector 84
norm of a vector in R3 84
Normal vector 117
Normal vectors 119
Note on absolute value 27
Note on absolute value inequalities 48
Note on addition and subtraction 29
Note on commutative and associative properties 22
Note on distributive property 20
Note on first-degree equations 38
Note on graphical representation 50
Note on higher-degree equations 42
Note on intervals 23
Note on multiplication and division 30
Note on open and closed intervals 25
Note on powers and roots 33
Note on radical equations 47
Note on radicals 44
Note on second-degree equations 40
Note on simplification and rationalization . 45
Note on solving inequalities 49
Nth root of a complex number 76

O

Odd function 166
Operations with functions 149, 151
opposite vector 87
order restriction 221
Orientation of the parabola 133, 135
Outward scaling 172

P

Parabola 132, 134, 136
Parabolic mirror 136
Parabolic trajectory 136
Parallel lines 106, 113
parallel lines 97
Parallel planes 119
parallel planes 100
Parallel vectors 112

ÍNDICE ALFABÉTICO

parallel vectors 96
Parallelism 113
Parallelism condition 112, 113
Parallelism of lines 114
Parametric equation 108
parametric equation of a line 96
Parametric equation of the intersection . . . 122
Parametrized intersection of planes 122
Perpendicular line 109
Perpendicular lines 106
perpendicular lines 98
Perpendicular planes 119
perpendicular planes 100
Perpendicular vectors 112
Perpendicularity in $\mathbb{R}^3$ 113
Perpendicularity of planes 121
Piecewise continuous model 158
Piecewise function 154
Piecewise functions in real-world models . 158
Piecewise-defined function 156
Plane 117
Plane equation passing through the origin 118
point belonging to a plane 99
Point on the circle 127
Point on the line 108
Point on the plane 117
Point-normal form of the plane equation . 118
Point-slope form 110
Power 31
power property of logarithms 206
Powers of Complex Numbers 74
product of matrix and adjoint 230
Product rule 151
Projectile trajectory equation 136
Properties of derivatives 149, 151
properties of logarithms 205
Properties of operations with functions ... 152
properties of the dot product 92
properties of vector addition 86
Property of the inverse function 196
property of the inverse matrix 232
Proportionality of area with radius 132
Pythagorean identity 212, 213
Pythagorean theorem 215

Q

Quadratic function composition 179
Quotient rule 151

R

Radical 42
Radical equation 46
Radical function 147, 148
Radius 127
Radius of the circle 129
Range 145, 148
Range of a function 148
Range of composite functions 146, 148
Rational function 147
Rationalization of radicals 44
Real-world example 158
rectangular form 67
Rectangular form of a complex number 57
Reflection 165, 167
Reflection of a linear function 166
Reflection of the graph of the inverse function 195
Reflection with respect to the x-axis 165
Reflection with respect to the y-axis 165
Reflective properties of the ellipse 141
Relationship between function and inverse 193, 196
relationship between tangent, sine, and cosine 212
Remark on Applications of nth Roots of Complex Numbers 78
Remark on De Moivre's Formula 73
Remark on graphical representation in the Argand plane 62
Remark on modulus and argument of complex numbers 65
Remark on rectangular form of complex numbers 59
Remark on the Application of De Moivre's Formula 76
Remark on the conjugate of a complex number 71
Right triangle 115
right triangle 212
Root 31
Rotational symmetry 166
Row Reduction 252

S

scalar multiplication 88
Scaling 170
Scaling and domain 170
Scaling and periodic functions 175
Scaling in graphs 175
Scaling of trigonometric functions 172
Scaling quadratic functions 174
secant 213
Second-degree equation 39
Semi-major axis 137
Semi-minor axis 137
Simplification of radicals 44
Simplified plane equation 119
singular matrix 228, 231
Sinusoidal composition 184
Sinusoidal function composition 179
Slope 113
Slope of a line 106
slope of a line 97
Slope-intercept form 110
Solving inequalities 49
solving systems using the inverse matrix . 236
Square root and composition 179
stability of nonlinear systems 258
Standard form of the circle 129
Standard form of the parabola 132
substitution in nonlinear systems 256
Substitution Method 243
Substitution Method, Matrix Method 248
Subtraction 27
subtraction of complex numbers 65
Subtraction of functions 149
Sum of distances in the ellipse 141
Surjective function 146
Symmetry of graph of a function and its inverse 195
Symmetry of inverse functions 194
Symmetry with respect to the y-axis 165
System of Equations, Substitution Method, Example 243

T

Temperature conversion with inverse functions 196
Theorem for graphical representation in two variables 50
Theorem of commutative and associative properties 21
Theorem of differentiability of composite functions 178
Theorem of distributive property 19
Theorem of equivalence of first-degree equations 38
Theorem of function reflection 166
Theorem of graphical transformations 168
Theorem of plane intersection 122
Theorem of radicals and powers 43
Theorem of solving absolute value inequalities 47
Theorem of solving linear inequalities 49
Theorem of the quadratic formula 39
Theorem on argument of a complex number 61
Theorem on distribution of subtraction ... 28
Theorem on extraneous solutions 46
Theorem on fraction simplification 30
Theorem on fractional powers 32
Theorem on function scaling 170
Theorem on horizontal scaling 173
Theorem on interval inclusion 24
Theorem on interval intersection 23
Theorem on multiplication of complex numbers 58
Theorem on polar form 63
Theorem on Powers of the Product of Complex Numbers 75
Theorem on rationalization of radicals 45
Theorem on roots using De Moivre's Formula 72
Theorem on scaling in graphs 175
theorem on solving exponential equations 208
Theorem on the conjugate of a product ... 70
Theorem on the distribution of nth roots ... 77
Theorem on triangle inequality 26
Theorem, Consistent Linear Systems 252
Theorem, Dependent and Inconsistent Systems 247, 250
theorem, homogeneous system 237
Theorem, Inconsistent Systems 246
Transformation of trigonometric functions 169
Translation 167
Translation theorem for functions 164
transpose of a matrix 224
transpose of matrix product 225

Transverse axis 139
trigonometric functions 211
trigonometric identity 214

U

Uniform circular motion 131
uniqueness of equality of exponents 208
uniqueness of equality of logarithms 209
Uniqueness of inverse functions 188
uniqueness of solution, lemma 236
Upward scaling 170
use of logarithms to solve exponential equations 209

V

vector 83
vector addition 86
vector addition and subtraction 86
Vector equation 108
vector in R2 83
vector subtraction 86
Verification of inverse functions 192
Verification of the inverse function 190
Vertex of the parabola 132
Vertex-focus distance 134
Vertical translation 163

Z

zero matrix 221
zero vector 87

BASIC MATHEMATICS

MBA. Helbert Justo Luque Zevallos

YEAR 2024

First Edition
ISBN:9798301465598

Series: Bachelor's Degree in Mathematics

- Basic Mathematics
- Logical Mathematical Reasoning
- Mathematical Analysis I
- Algebra
- Statistics and Probability
- Mathematical Analysis II
- Linear Algebra I
- Statistical Inference
- Real Analysis I
- Numerical Analysis
- Linear Algebra II
- Algebraic Structures
- Topology
- Real Analysis II
- Ordinary Differential Equations
- Linear Optimization
- Partial Differential Equations
- Introduction to Hyperbolic Geometry
- Galois Theory
- Numerical Methods for Solving Differential Equations
- Measure and Integration
- Nonlinear Optimization
- Qualitative Theory
- Functional Analysis
- Differential Geometry I
- Introduction to Algebraic Topology
- Differentiable Manifolds
- Introduction to Variational Methods for Differential Equations
- Introduction to Differential Topology
- Minimal Surfaces I
- Differential Geometry II
- Introduction to the Finite Element Method
- Introduction to the Geometry of Differential Forms

www.ingramcontent.com/pod-product-compliance
Lightning Source LLC
Chambersburg PA
CBHW082243220526
45469CB00009B/2867